environmental
quality
by design: south florida

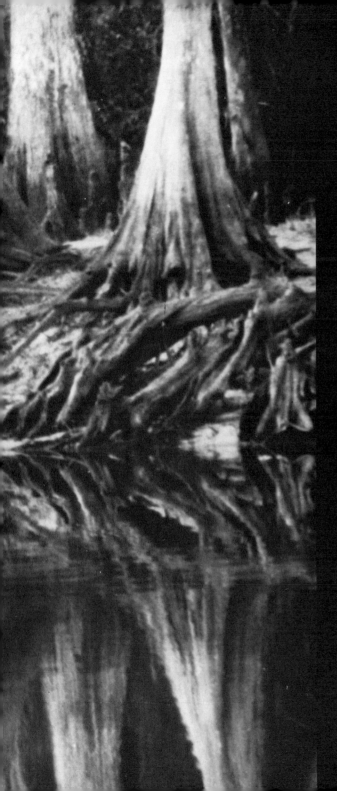

environmental quality by design: south florida

Albert R. Veri,
William W. Jenna, Jr.,
Dorothy Eden Bergamaschi

with cartography by
Linda C. Sumarlidason

In cooperation with the
Center for Urban and Regional Studies,
University of Miami

University of Miami Press
Coral Gables, Florida

ABOUT THE AUTHORS

Albert R. Veri is a land planner and landscape architect with Albert R. Veri and Associates. All three of the authors are associated with the environmental programs at the Center for Urban and Regional Studies at the University of Miami.

William W. Jenna, Jr., is the author of two demographic studies, *Metropolitan Miami* and *Metropolitan Broward*, published jointly by the Center for Urban and Regional Studies and the University of Miami Press.

Dorothy Bergamaschi is a landscape architect and urban and regional planner in private practice.

Consulting biologist: Susan Uhl Wilson

Environmental law consultant: Timothy Keyser

Illustration credits:
All photographs by Brian Bergamaschi except as noted: Hernando Acosta, p. 147, right; Yiannis Antoniandis, p. 146, second from left; Ted Baker, p. 146, third from left; H. Carlton Decker, p. 147, left; Dillon Aerial Survey, p. 48; *Miami Herald*, p. 46, third from left, p. 62, p. 74, p. 129; Miami Metro News Bureau, p. 86; National Park Bureau, p. 30; unknown, p. 108.

Line drawings and diagrams by the following: Adonay Bergamaschi, p. 84, p. 126; Dorothy Bergamaschi, pp. 72-73, p. 76, p. 77 left, pp. 80-81, pp. 88-89, pp. 104-5, pp. 112-13, pp. 120-21, p. 127; Susan Stevens Suarez, p. 69, p. 75, p. 77 right, p. 78, p. 79, p. 87, pp. 100-101, p. 140, p. 153, p. 154, p. 156, p. 157, p. 158, p. 159, p. 160; Linda Sumarlidason, pp. 10-67, p. 152; Albert Veri, p. 133, p. 139, p. 141, p. 145, pp. 148-49.

This work was sponsored by a grant from The Construction Industry Advancement Fund, administered by The Associated General Contractors, South Florida Chapter.

Manufactured in the United States of America

Library of Congress Cataloging in Publication Data

Veri, Albert R   1940-
    Environmental quality by design, South Florida.

    "In cooperation with the Center for Urban and Regional Studies, University of Miami."
    Bibliography: p.
    Includes index.
    1. Environmental protection—Florida.   2. Environmental policy—Florida.   I. Jenna, William W., joint author.   II. Bergamaschi, Dorothy Eden, 1934-   joint author.   III. Sumarlidason, Linda C.   IV. Miami, University of, Coral Gables, Fla. Center for Urban and Regional Studies.   V. Title.
TD171.3.F6V47   363.6   75-26950
ISBN 0-87024-279-2

contents

part one: overview   13

introduction □ the south florida region □ florida emerges □ geology and soils □ vegetation □ climate □ water resources □ consumption □ flood and storm □ wildlife □ ecological region □ land use □ demography □ infrastructure □ recreation □ florida's past

part two: the natural environment   65

the natural environment □ the coastal strip □ the florida keys □ the sandy flatlands □ the interior wetlands □ mangrove and coastal marsh

part three: the built environment   129

the built environment □ urban region □ water: a regional issue □ land use and the law

figures

tables

foreword

With the spirit and conviction that each individual must diligently work to change a small portion of events to leave a legacy of improved communal living, the Construction Industry Advancement Program, administered by the South Florida Chapter of the Associated General Contractors, is pleased to sponsor this book.

The rapid growth, rich heritage, and enormous potential of the state of Florida should induce Floridians to renew their individual and collective dedication to work together for a better state. If the past is prologue to the future, then the coming years surely promise more expansion, more people, and more industry in Florida. This future belongs to the heirs of our state, whoever they may be, yet we must ask ourselves what will be the nature of the legacy we leave. Will it be total preservation of the environment or will it be total development? Obviously, neither alternative is a sensible solution.

This book demonstrates the compatibility between good, sound growth and development planning and good, sound environmental practices. Although most of us are basically environmentally minded, the issue that we each must resolve is whether or not we will adhere unrelentingly to our own personal preferences or whether we will be willing to work within a positive framework toward an end where the community is enhanced for all of its members.

If Florida is to prosper, environmentalists and growth entrepreneurs must meet in an atmosphere of understanding and mutual cooperation. A spirit of free exchange, a sense of give and take, must exist on both sides of the "growth versus environment" question. We cannot allow the free enterprise system—the "American Way"—to fall into a state of chaos because of the lack of concern by reasonable men, nor can we continue to make one-sided decisions.

Bringing together diverse elements of a rapidly expanding society is not easily or quickly accomplished. In this tedious process, time is both an ally and an enemy. More time will provide more interchange and greater understanding; however, because it is important to move with haste toward the goal we have set for ourselves —that of a quality environment for all—we must act now to identify the problems in order to resolve them.

To provide insight and information, the Associated General Contractors through the Construction Industry Advancement Program sponsored this book with a grant to the Center for Urban and Regional Studies at the University of Miami. It is the belief of the Construction Industry Advancement Program that accurate information, such as is presented in this volume, will contribute to an alliance between the environmentalists and the members of the construction industry. Such cooperative efforts will lead to the realization that growth and environmental protection can work hand in hand to build a dynamic and progressive future. The end result will be a better life for us all.

8

acknowl-edgments

This book has been in preparation for more than eighteen months, and throughout this entire period we have made every effort to obtain and incorporate a spectrum of ideas from individuals including government officials, engineers, architects, scientists, informed citizens, and representatives of the construction industry. As a result, we hope that this book will serve as a useful tool in the coming years when critical decisions will be made regarding the continuing development of southern Florida.

We wish to express our gratitude to the South Florida Chapter of the Associated General Contractors who sponsored the writing and publication of this work. Beyond their financial backing, however, we appreciated their technical knowledge and constant encouragement.

Special thanks are extended to Dr. Carl E. B. McKenry, Vice President for Academic Affairs at the University of Miami, whose efforts resulted in the creation of the Center for Urban and Regional Studies, under whose auspices this study was produced, and to the Interim Director of the Center, Priscilla Perry, who facilitated interdepartmental involvement and provided assistance to our study team.

We wish to acknowledge the following persons who consulted with us and provided new insights. Their names are listed alphabetically: Hernando Acosta, Yiannis Antoniadis, Ted Baker, Adonay Bergamaschi, Mr. and Mrs. Sam Clark, H. Carlton Decker, Shakima De Jesus, Marjorie Stoneman Douglas, Ivan Ficken, William Forney, John A. Fortes, Harold P. Gerrish, Alan Gold, Dr. Seymour Goldweber, Dr. Leonard Greenfield, James Hartwell, Dr. Eric Heald, E. T. Heinen, Terry Holzheimer, John Hope, Suzanne Jenna, Antonio Jurado, Howard Klein, Jay Landers, Carol Levy, Dr. Benjamin McPherson, Frederick F. Monroe, Joseph Nash, Dr. Dennis O'Connor, Dr. Oscar Owre, M. Barry Peterson, Maria Posada, James H. Sayes, M. Sally Schauman, Admiral Orvan R. Smeder, Dr. William M. Stall, Earl M. Starnes, Dr. Durbin Tabb, Glenn Taylor, Dr. Charlton W. Tebeau, Eastern Tin, Earl Van Atta.

To all these very special people, we extend our heartfelt thanks.

9

introduction

1975

Count Alexis de Tocqueville in 1835 prefaced his great treatises on American life, customs, manners, and politics with the statement: "We should not judge the future by the ways of the past." However, in unprecedented numbers American builders and developers are not following such advice at a time when this admonition is particularly apt and critical. Nationwide, literally millions of acres of land have been bought with the idea of development, indeed a commitment to develop, which draws upon old ideas and old ways even though there are ample signs of major changes taking place.

In the last 10 years the United States witnessed a new resurgence of an old mania that surpasses even the westward rush of frontier days in America—the frenzy to acquire land. Nowhere in the country has this trend been seen more clearly than in southern Florida, which last year led the nation in the rate of new construction, topping even such giant metropolitan regions as New York, Chicago, and Los Angeles.

Everyone must now assume a greater responsibility than ever before, particu-larly in an area as environmentally fragile as southern Florida. We believe environmental consciousness in the planning and production of the built environment will serve the best social-economic-environmental interests of the community at large. In short, we must not create new problems but rather we must solve existing ones.

The use of the land in southern Florida began with both courage and ignorance: courage to inhabit a land that appeared to favor fish and wildlife, and ignorance that destroyed parts of the richness of the region by indiscriminate clearing, draining, and filling. It is reasonable to believe that there is no area in southern Florida in which it is technically impossible to build nor beyond the financial reach of most consumers. Technology makes it possible to modify any area. However, if we fail to recognize the costs as well as the benefits that major modifications have upon the total community, we can be likened to the proverbial ostrich with his head in the sand. One can point to numerous examples of where the developer and the consumer have paid the price for their apathy—pollution, traffic congestion, water shortages, building moratoriums, and many more problems too numerous to list. Unfortunately this situation tends to overshadow the good we have accomplished in a high standard of living, available housing, mobility, a free government. Although perfection may never be achieved, each effort that we make should contribute to the total effort and should bring us closer to our goal. A satisfactory quality of life for all is not incompatible with a growing and stable economy.

The idiom of "Who is more important, man or birds?" is literally an oversimplification of the realities of what is at stake. Man must exist in conscious harmony with nature; there is no other choice. Cutting against the grain of natural systems is usually more expensive in terms of construction and maintenance costs, and it is disastrous in terms of the loss of social amenities and political effectiveness. Society has neither the resources nor the financial liquidity to bankroll a synthetic environment. Time, money, and energy spent fighting a particular issue could more reasonably be spent improving the quality of the built environment.

The purpose of this book is to lend insight to the issues as they exist, with the hope that creative minds working toward common goals will significantly enhance environmental quality by design.

We are grateful to the Construction Industry Advancement Fund, sponsored and administered by the South Florida Chapter of the Associated General Contractors, which perceived the timeliness for this publication.

The information is presented in three parts:

Part One is an overview of the southern Florida region, its natural resources, social

FIGURE 1: Map showing study area of Caribbean.

and economic characteristics, and their geographic distribution. Its purpose is to provide a survey of the richness and the diversity of southern Florida.

Part Two examines each of the ecologic regions of southern Florida, their structural and dynamic characteristics, and in particular their function in supporting human life and ecosystems upon which human existence depends.

Part Three is a compendium of recommendations with cross references to Parts One and Two. This section deals with the planning process, principles and policies for resource management and building, and land use and the law.

A. R. V.
W. W. J.
D. E. B.

June, 1975

the 2
south florida region

Map of the south Florida region showing counties (CHARLOTTE, GLADES, MARTIN, LEE, HENDRY, PALM BEACH, COLLIER, BROWARD, MONROE, DADE), cities, LAKE OKEECHOBEE, Big Cypress Swamp, Everglades National Park, Conservation Areas 1, 2, 3, and geographic features including the GULF OF MEXICO, ATLANTIC OCEAN, FLORIDA BAY, and the FLORIDA KEYS.

STUART
MARTIN
CHARLOTTE
GLADES
LAKE OKEECHOBEE
PUNTA GORDA
MOORE HAVEN
PALM BEACH
LA BELLE
LEE
HENDRY
FORT MYERS
PALM BEACH
SANIBEL ISLAND
Immokalee
Area 2
1
COLLIER
Alligator Alley
State Road 84
Conservation
BROWARD
FORT LAUDERDALE
NAPLES
Big Cypress Swamp
3
MARCO ISLAND
Tamiami Trail
TEN THOUSAND ISLANDS
(US 41)
MIAMI
KEY BISCAYNE
GULF OF MEXICO
DADE
MONROE
Everglades National Park
ELLIOTT KEY
Dixie Hwy.
FLORIDA BAY
ATLANTIC OCEAN
FLORIDA KEYS
KEY WEST
I-95
US 1

north
mi. 0 5 10 20 30 40
km. 0 5 10 20 30 40 50 60

part one

overview **1**

florida emerges

Peninsular Florida is the emergent part of the Floridan Plateau, a projection of the continental land mass of North America which separates the deep waters of the Atlantic from the Gulf of Mexico. South of Palm Beach the eastern edge of the plateau lies only two to three miles off the Atlantic coast, sweeping close to the arc of the Florida Keys and past the Tortugas. In the Gulf of Mexico, the plateau extends more than 100 miles beyond the coast.

Alternating periods of high and low water created the landforms of southern Florida. During periods of submergence, layers of limestone were deposited, smoothed, and troweled by the action of the sea.

Each successive sea level left its mark, and the beach terraces and scarps that marked the ancient shorelines can still be traced. Today the sea level is rising again at a rate of three inches per 100 years, and land building processes are active along the shore and on the ocean floor. (Parker and Cooke, 1944.)

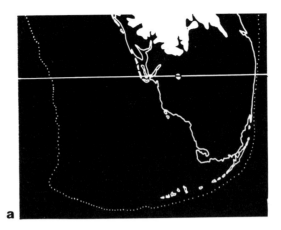

a

Talbot Shoreline

During the Sangamon interglacial period the sea stood 42 feet above its present level (present level is shown as 0 in Figure 3b) and all of southern Florida was submerged except for Immokalee Island. Miami limestone was being deposited in the southeast, while the shell beds and sandy limestone of the Anastasia formation were accumulating as offshore bars. Coral reefs were flourishing at the southeastern edge of the plateau. These reefs were to become the Florida Keys.

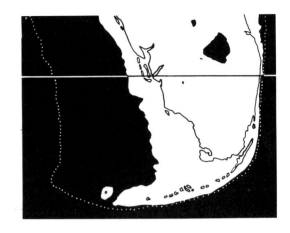

Shoreline in Iowan Times

Erosion was active during the Iowan glacial substage, when the sea fell to 60 feet below its present level. Transverse channels were cut through the coastal ridge by the erosive force of the outflowing freshwater. The higher portions of the limestone ridge were pitted by solution holes, and sand transported by ocean currents built bars and dunes along the shore.

b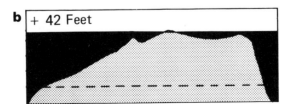

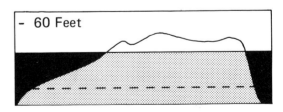

FIGURE 3: Depicts the sequent shorelines of Florida. The edge of the Floridan Plateau, 300-feet below present mean sea level, is shown dotted. The sections below compare the prehistoric sea elevations to present sea elevations. (Adapted from Parker and Cooke, 1944.)

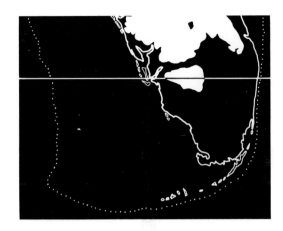

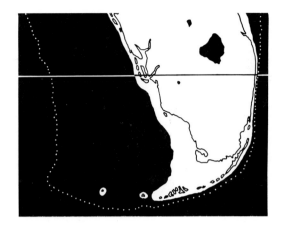

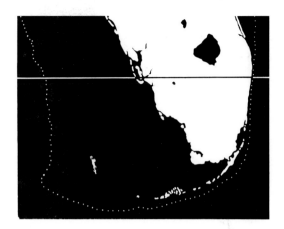

Pamlico Shoreline

Southern Florida was flooded again as the sea rose to the 25-foot level in Pamlico times. Lake Okeechobee became a bay, and Immokalee was an island again but larger than that of the Sangamon Period. Quartz sands, still being transported from the north, were distributed over the earlier formations as far south as Miami, as well as in the northern part of the Big Cypress, forming high dunes in some places.

Late Wisconsin

The slow retreat of the sea to minus 25 feet was halted periodically, forming a series of parallel shorelines. These fossil beach ridges and dunes can be seen in many parts of southern Florida, particularly in Martin and Palm Beach counties on the east coast, and at Marco and near the town of Everglades to the west. Limestone erosion resumed, and Lake Flirt marl was deposited near the Caloosahatchee River and in the Everglades.

Recent Times

As the glaciers retreated for the last time, the sea slowly rose to its present level, flooding the ends of existing streams to form modern bays and inlets. The tidelands of the Ten Thousand Islands, now separated from the mainland, were invaded by mangroves. Saw grass began to grow and form the peaty soils of the Everglades, while Lake Flirt marl continued to be deposited. Wave-borne sand and shell formed modern beaches and dunes on the mainland; in the clear waters of the Keys a new coral reef began to grow.

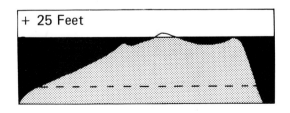

+ 25 Feet

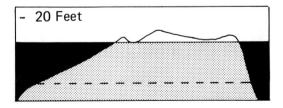

− 20 Feet

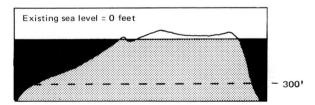

Existing sea level = 0 feet

− 300'

geology & soils

"Limestone" is the key word in Florida geology. There are 800 miles of it extending from Key West to Tallahassee, in layers 12,000 feet thick, reaching to the granite base of the continent. According to thorium dating, the main body was formed about 190,000 years ago, during periods of geological time when the Florida peninsula was below sea level and the processes of limestone deposition were active. At the present time these same processes are creating new limestone deposits in the seas around our shores.

Limestone is composed of calcite (calcium carbonate or $CaCO_3$). This material is deposited in various ways, each producing a different type of limestone.

1. **Oolitic Limestone:** Oolite is formed in shallow mud seas by precipitation of $CaCO_3$ from the ocean water around a small nucleus, which can be a piece of shell or other material. The precipitate forms onion-like layers. The pinhead-sized round particles then become cemented together to form a limestone.

2. **Bryozoan Limestone:** This type is formed by Bryozoan animals similar to the polyps that form coral. $CaCO_3$ is precipitated from the ocean water to form the shell and is simultaneously excreted into the surrounding mud, thereby creating a marl deposit.

3. **Coral Reef Limestone** (Key Largo Limestone): This formation is found in the Florida Keys and is the only recent coral limestone formation in the United States. In the Keys area the conditions necessary for coral formation exist. These are: (a) clear, unpolluted water (sunlight must penetrate); (b) existence of algae to supply oxygen; (c) water less than 150 feet deep (again so that sunlight may penetrate to support algal growth); and (d) water temperature above 70° but not more than 90°. The coral animal precipitates calcium carbonate from the ocean water to form its supporting structure.

4. **Coquina Limestone:** Coquina limestone is found on the west coast of Florida where shell animals are plentiful and the sand is composed mostly of broken shells. The existence of shellfish depends on the plentiful nutrients available in these waters because of upwelling offshore currents. These animals form their shells by precipitating calcium carbonate from the water, and, when they die and the shells become broken through pressure or wave action, the fragments become consolidated into coquina limestone.

5. **Marl:** Marl may be formed by excretion of $CaCO_3$ by marine organisms in salt water, or in freshwater by precipitation through the action of the blue-green algal mat that covers the ground in wetlands. It is a fine-textured, clayey deposit that can become hardened over time into a dense limestone. Unlike the other limestones, its lack of pore spaces makes it relatively impervious.

In most geological areas, rivers, winds,

and glaciation are the major mechanisms of erosion. None of these is of importance in South Florida. Here the dominant erosional agent is underground water, which carries away dissolved limestone in solution. The chemical process begins in the atmosphere, which is composed of oxygen, nitrogen, and other gases, one of which is carbon dioxide (CO_2). As rain falls, it absorbs CO_2 from the air to form a weak acid that percolates through pore spaces and cracks in the limestone, slowly dissolving the rock. One foot of limestone is dissolved each 800 to 1,000 years, which is a rapid rate of erosion. This process created the famous Mammoth Cave of Kentucky, as well as the numerous lakes and sinkholes that are a striking feature of the South Florida landscape. The water, now "hard" because of the dissolved limestone ($CaCO_3$) it contains, flows underground until it reaches the sea. There, limestone deposition occurs again, completing the cycle.

Surficial Deposits

Figure 4 illustrates the geological formation below the region's land surface. The general distribution of surficial sands and organic soils are illustrated in Figure 5. The surficial deposits are, in fact, the "soils" of this region.

The surficial sand deposited during the Pamlico Period is usually highly permeable; Talbot and Penholoway sands are less permeable, and the organic soils and marls are the least permeable. Pervious surface deposits overlying pervious geologic formations are excellent conductors for groundwater recharge. Impervious surficial deposits severely limit percolation to the geological formation with resultant heavy runoff of waters falling on the surface. In the case of organic soils, however, much water is absorbed to the point of saturation, before runoff occurs. Figure 5 illustrates this association. Because of diversity of location, these associations are difficult to map at the regional scale, and, therefore, proper evaluation requires a careful on-site inspection.

Influences Upon Construction

An inventory of the geology and soils in southern Florida is summarized in Table 1. The 200-foot depth (Figure 7) ranges from surface marls through to the permeable part of the Tamiami formation, an aquiclude forming the bottom layer of the Biscayne Aquifer.

This section describes the geological influence upon construction as it relates to:

1. **Characteristics:** Physical formation, granular content, and weight-bearing capacity.

2. **Thickness:** Relative thickness of the formation (see Figure 7).

3. **Aquifer Potential:** Potable water bearing capacity, important in siting and determination of potential water supply (see Figure 18).

Lake Flirt Marl is an impervious formation, with poor aquifer potential. In places it is hardened to a dense formation, one to six feet thick. Rainfall usually runs off to lower elevations, turning sinkholes into natural lakes. Due to its high impermeability, marl permits little groundwater recharge; where it underlies the very pervious Pamlico sands, it may support a perched water table (Table 4).

The bearing capacity of the marls varies with thickness, although there is minor shrink-swell potential. In lower elevations, the surface area is usually flooded. Since no percolation occurs, drainage is usually necessary before the site can be used for farming or building.

Pamlico Sands rarely occur above 25 feet mean sea level (MSL) except in the case of dunes or ridges where sand is heaped into thick deposits. Usually the sand is only one or two feet thick. Often the white quartz sands are stained black or red by eroded sedimentary materials or iron oxide. They usually overlie the Anastasia or Miami Oolite formations.

Water-bearing capacity varies with compaction. Many wells are required to obtain significant quantities of water from these sands. Bearing capacity varies with location and proposed land use. In those areas where depth is thin (1 to 2 feet) the underlying formation determines bearing capacity limits.

When stripped of vegetation, the Pamlico sands are shifted easily by wind and eroded by water. Extensive erosion control measures are required for major developments where land clearing is involved.

Miami Limestone, a soft, white to yellow limestone formation containing streaks of calcite, is located in Broward, Dade, and the Everglades portion of Monroe counties (Figure 4). Varying in depths from 1 to 40 feet, the formation has an excellent water-bearing capacity, although yields are somewhat low because it quickly loses its water. This shallow aquifer yields high-quality water in areas which have not been contaminated by polluted runoff or groundwater. Its greatest yields occur where it meets the Fort Thompson formation, a highly productive aquifer.

Oolite is easily eroded, and subsurface solution erosion creates cavities that can result in damage to foundations and roadways. Below the surface, conditions are unpredictable, and pilings driven 10 to 20 feet apart may differ many feet in length.

Miami limestone forms the Atlantic Ridge south of Boca Raton. The elevation of this portion of the ridge ranges from 10 to 20 feet. The base of this formation is seldom 20 feet below sea level.

Anastasia Formation is the backbone of the Atlantic Ridge along the coast from Boca Raton to northwest of Lake Okeechobee and beyond. Another section reaches from Naples north to Tampa and westward along the Caloosahatchee River (Figure 4). This wedge-shaped formation is approximately 100 feet thick.

TABLE 1

SUMMARY OF SOUTH FLORIDA GEOLOGICAL CHARACTERISTICS

(Source: Parker, 1943; Hoffmeister, 1974)

FORMATION	PHYSICAL CHARACTERISTICS	*THICKNESS IN FEET	AQUIFER POTENTIAL
Lake Flirt Marl	Relatively impermeable Dense limestone (white to gray calcareous mud rich with shells)	0 to 6	Poor
Pamlico Sands	Highly permeable Quartz sand (white, black, or red)	0 to 60	Low yields
High Terrace Deposits	Generally permeable Quartz sand with intercalated clay and silt beds (generally unconsolidated, but consolidated locally to sandstone)	0 to 100	Low yields
Miami Limestone	High to good permeability Soft, white to yellow limestone Solution erosion causes cavities in the formation	0 to 40	Excellent, but low water holding capacity
Anastasia Formation	High to fair permeability Sand, sandy limestone, shell marl, coquina	0 to 100	Good to fair yields to wells
Key Largo Limestone	Highly permeable Coralline reef rock, ranging from hard and dense to soft and cavernous	0 to 60	Excellent, but easily intruded by seawater
Fort Thompson Formation	Permeability: low in northern Everglades, high in the south Alternating marine, brackish and freshwater marls, limestone, and sandstones	0 to 200	Excellent in south—forms major part of Biscayne Aquifer
Caloosahatchee Marl	Low permeability Sandy marl, clay, silt, sand, and shell beds	0 to 50	Fair to poor
Tamiami Limestone	High permeability in upper part Lower portion forms aquiclude at base of Biscayne Aquifer Creamy white limestone, greenish gray marl, silty and shelly sands	0 to 150	Excellent in upper portion Forms lower part of Biscayne Aquifer

*See Figure 7 for generalized depth characteristics

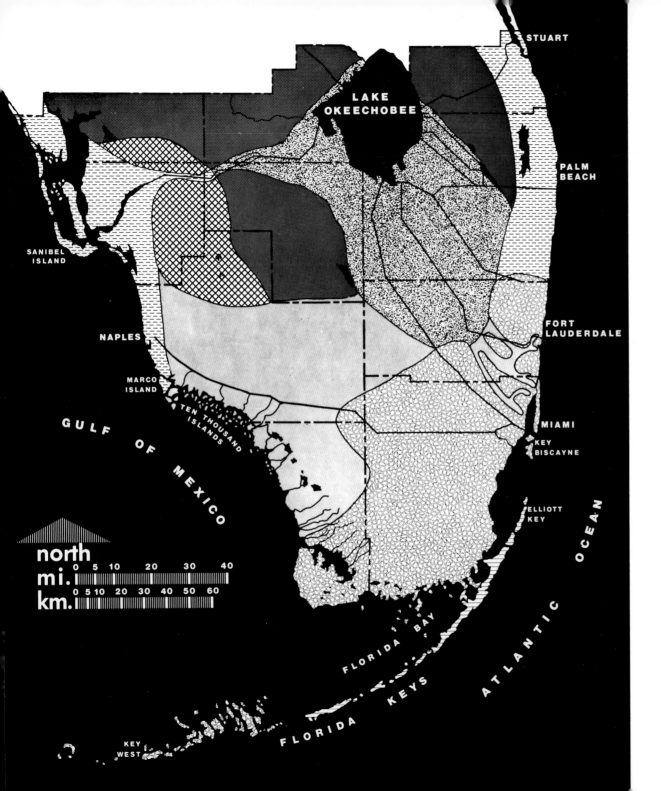

STUART

LAKE OKEECHOBEE

PALM BEACH

SANIBEL ISLAND

NAPLES

FORT LAUDERDALE

MARCO ISLAND

TEN THOUSAND ISLANDS

GULF OF MEXICO

MIAMI

KEY BISCAYNE

ELLIOTT KEY

north
mi. 0 5 10 20 30 40
km. 0 5 10 20 30 40 50 60

FLORIDA BAY

KEY WEST

FLORIDA KEYS

ATLANTIC OCEAN

geology 4

◻ **LAKE FLIRT MARL**
Chiefly soft freshwater marl, locally case-hardened. Found in patches throughout the Everglades.

▦ **MIAMI LIMESTONE**
Oolitic and Bryozoan limestone, sandy where it grades into the Anastasia Formation.

▤ **ANASTASIA FORMATION**
Coquina, sandy limestone, shell, marl, sand.

▤ **KEY LARGO LIMESTONE**
Coral reef limestone.

▓ **FORT THOMPSON FORMATION**
Alternating marine, brackish, and freshwater marls, limestones, and shell beds. Uppermost portions are contemporaneous with Anastasia, Miami Oolite, and Key Largo Formations.

▦ **TAMIAMI FORMATION**
Calcareous sandstone, sandy limestone, and beds and pockets of quartz sand. Permeability generally very high. Main component of the Biscayne Aquifer.

■ **CALOOSAHATCHEE MARL**
Gray, sandy to calcareous shell marl.

▩ **BUCKINGHAM MARL**
Clayey and calcareous marl, locally case-hardened to brittle limestone. Contains considerable amount of phosphate sand and pebbles.

SOURCES

Florida Geological Survey, Bulletin 27, Plate 15 (Geologic Map of Southern Florida Exclusive of Surficial Sands), Late Cenozoic Geology of Southern Florida, with a discussion of the groundwater, by Garald G. Parker and C. Wythe Cooke (U.S.G.S.) Tallahassee, Florida (1944).

The deposits of this formation are as old as the marine members of the Fort Thompson formation. Classified as fair to highly permeable, the Anastasia formation's aquifer potential is rated as good; however, canals penetrating the formation will lower the water table. In general, bearing capacity is good, but susceptible to underground erosion.

Key Largo Limestone includes parts of a dead coral reef forming the coastal islands from Soldier Key south (see Figure 4). This formation is three miles wide, 90 miles long, and 60 feet deep at sea level. Its base is thought to be wider and longer. Its permeability is high, but water-bearing capacity is low because, being highly porous, it readily leaks to the sea. In the Keys, the highly saline groundwater is used only for fire fighting or flushing.

Composed of coralline reef rock, ranging from hard and dense to soft and cavernous, this formation was cut into blocks and slabs for building by the early settlers. On the ocean side, the formation is exposed at the surface; but, on the protected Florida Bay side, overlying sands and sediment support a mangrove forest.

Fort Thompson Formation varies in permeability, increasing in water-bearing capacity from north to south. In the northern part of the region where it underlies Lake Okeechobee, its impermeability enables the 717 square-mile lake to hold water. In this region the formation averages eight feet thick. Toward the south, in Palm Beach, Broward, and Dade counties, it forms part of the Biscayne Aquifer, which provides excellent potable water for the "Gold Coast." At the latitude of Miami, the formation ranges from 80 to 200 feet thick. It is underlain throughout by the Tamiami formation, which serves as an aquiclude.

The Fort Thompson formation also varies in its physical characteristics from north to south, with alternating marine, brackish, and freshwater marls in the north-central region and limestones and sandstones in the south.

Bearing capacity is variable and is most affected by underground caverns. Test borings are necessary for large projects throughout the study area to determine bearing capacity, and even then, small but deep caverns may be overlooked. In addition, solution holes are being formed constantly.

Caloosahatchee Marl Formation girdles the northern section of Lake Okeechobee and is found throughout the region as erosion remnants. Less than 50 feet thick, the formation underlies much of the Everglades and Big Cypress watershed.

The water-bearing capacity of the formation is poor to fair, yielding hard water near the coast in permeable areas. South of Lake Okeechobee and in the Everglades, water is highly mineralized and is unsuitable for drinking unless extensively treated. Aquifer potential in the western portion of South Florida ranges from poor to fair.

Tamiami Formation is permeable in its upper portion, and forms the lower part of the Biscayne Aquifer. The lower, and major portion of the Tamiami, forms the Floridan Aquiclude, sealing off the upper freshwater-bearing strata of the aquifer from the brackish water below. The Tamiami formation lies near the surface in Collier, Lee, and Charlotte counties (Figure 7).

SOILS

The soils of southern Florida are generally shallow and geologically immature in comparison to similar soils in the rest of the state. Because of high rainfall and the lack of topographic relief which would facilitate natural drainage, the physical development of the soils of southern Florida has been somewhat slow in a geologic time sense. These shallow soils are immaturely profiled and infertile when cleared of vegetation.

The warm subtropical climate and rainfall leach nutrients from the exposed sands, marls, and rock land. The slightly more fertile bog soils (mucks and peats) remained mostly underwater until this century when extensive drainage projects lowered the water table, exposing them to rapid oxidation.

New soils are created by decaying plants and by sediments trapped by roots and stems. These new soils and the modification of existing soils are, in places,

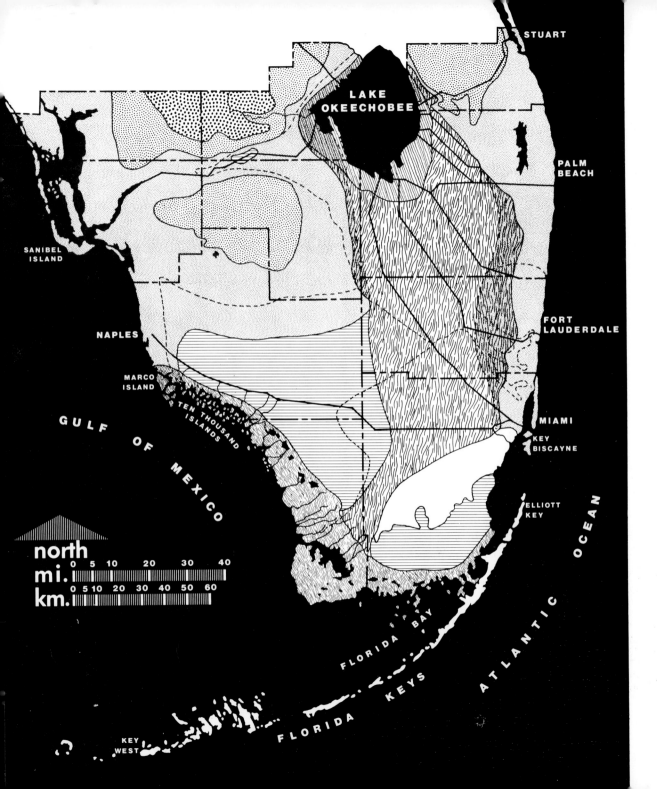

surficial deposits 5

 LAKE OKEECHOBEE MUCK
Organic, marly soils located in the peripheral marshlands around Lake Okeechobee.

 EVERGLADES PEATS
Organic soils of peat and muck with widespread deposits of calcareous marl. Almost a flat surface and generally wet throughout the year except where drained.

 MANGROVE PEAT
Very low, wet areas of organic, marly to mucky soils, thinly overlying bedrock.

 TALBOT FORMATION
Marine quartz sands locally hardened to form a sandstone. Found at elevations between 15 to 42 feet above mean sea level.

 PENHOLOWAY FORMATION
Marine quartz sands. Found at elevations between 42 to 70 feet above mean sea level.

 PAMLICO FORMATION
Marine quartz sands, in places hardened to a sandstone. Found at elevations up to 25' above mean sea level.

 Marl with fingers of peat

 Exposed limestone rock

 Extent of bedrock formations underlying surficial deposits.

SOURCES

Florida Geological Survey, Bulletin 27, Late Cenozoic Geology of Southern Florida with discussion of the groundwater by Garald G. Parker and C. Wythe Cooke. (U.S.G.S.) Tallahassee, Florida (1944); Plate 14: Surficial Deposits of Southern Florida Exclusive of Organic Soils.

Florida Geological Survey, Bulletin 30, Figure 13, p. 122: Isopach Map showing thickness of peat in the Everglades.

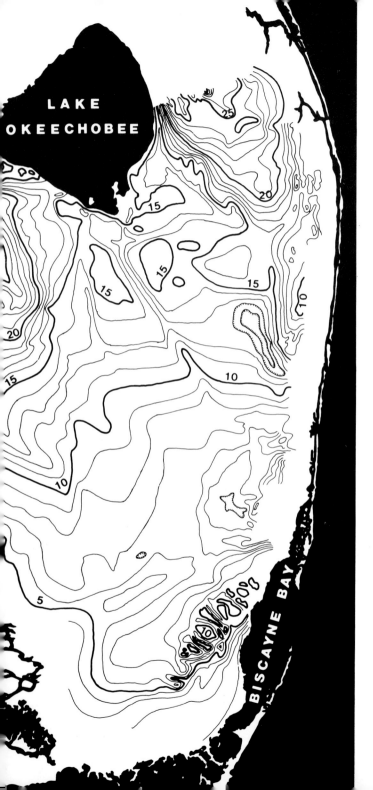

TABLE 2
SUMMARY OF SOUTH FLORIDA SOIL FORMATIONS

SOIL TYPE	BEARING CAPACITY	PERCOLATION	CONSTITUENTS	SHRINK-SWELL POTENTIAL
Rock (Limestone)	Good, but can contain cavities locally		$CaCO_3$ (calcium carbonate) from coral or other sources; cemented	Soluble in weak acids, cavities can develop, etc.
Sand	Good if confined	Rapid, excessive for some vegetation	Loose quartz grains, shell fragments	Highly stable
Marl	Deforms, shifts while wet (similar to clay)	Very slow, sometimes impervious	Fine particles of $CaCO_3$; a soft solid when dry, clayey when wet; locally hardened	Similar to clay
Organics (peat and muck)	Highly compressible, oxidize and disappear when dry	Varies	Partially decomposed plant remains	Extreme

FIGURE 6 depicts the bedrock contours below the major peat and muck areas. Because more extensive information is not available for the entire region, bedrock contours below the sand deposits are not shown. All data refer to mean sea level. (Source: Parker and Cooke, 1944).

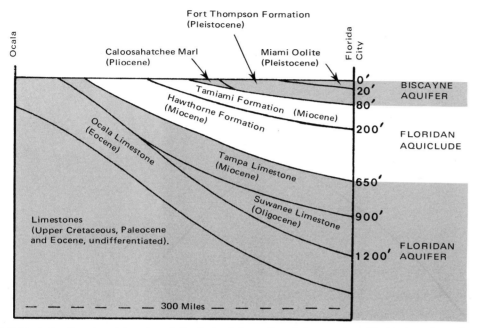

FIGURE 7 illustrates the general layering of the bedrock formations below southern Florida from Ocala (north) to Florida City (south). (Source: Parker, et al., 1955, p. 63).

TABLE 3 COUNTIES WITH PEAT DEPOSIT

COUNTY	SQUARE MILES	MAIN REGION AND AREAS
Monroe	145	Coastal marshes; swamps (mostly mainland)
Dade	625	Everglades; coastal marshes, swamps
Broward	700	Everglades, and a few coastal areas
Palm Beach	1,060	Everglades, and a few coastal areas
Martin	559	Everglades, and a few coastal areas
Collier	15	Corkscrew Swamp, parts of the Big Cypress, coastal marshes, and swamps
Hendry	165	Everglades, some marshy sloughs
Glades	65	Everglades, Fisheating Creek

creating new vegetation patterns (Alexander, 1973).

Peat is an accumulation of partly decomposed and disintegrated organic materials, derived mostly from plants. This material naturally accumulates where water abounds and varies in consistency from fibrous, matted, turflike material to a mudlike plastic ooze.

Muck is more completely broken down, and the plant fibers are no longer recognizable. It is usually granular to powdery and much darker in color than the peat from which it was derived. Much Everglades farmland is surfaced by muck.

Soils-Vegetation Association

For many years plant biologists have recognized the relationship between the importance of soil types and plant distribution. The vegetation of an area invariably gives a good clue to its soil characteristics. Almost 100 soil types have been analyzed and classified by the United States Soil Conservation Service and the Florida Agricultural Experiment Stations. For the purpose of describing the soil-vegetation association in Part One, we use the broader soil series—rocky soils, sandy soils, marl soils, and organic soils. Figure 9 shows the variety of vegetative communities occurring in southern Florida. Clearing, draining, agriculture, and community development have all contributed to changes in this vegetation pattern, which has been extensively modified in disturbed areas.

TABLE 4   DEGREE OF CONSTRAINT TO INTENSIVE DEVELOPMENT

The purpose of this table and the accompanying map is to illustrate broad regional constraints. Within the classifications shown there are smaller areas that may fall into other categories which cannot be shown at this scale.

	1	**2**	**3**	**4**	**5**
Vegetation	Pine-land	Pine flatwoods	Dry prairies Pineland	Wet Prairies Mixed cypress and pineland	Wetland formations Other critical areas
Soils	Sands	Sands Rock lands	Sands Rock lands	Sands Rock lands Marls	Organics Rock lands
Hydrology	Good Natural Drainage	Water table remains below surface, but after intense rain, standing water may persist for hours	Water table rises above surface for days during the rainy season causing widespread flooding	Water table rises above surface during the rainy season. Excess water feeds wetlands by surface flow. In deep sandy soils, surface water is absorbed to recharge shallow aquifers	Mainland areas flooded for most of the year. Areas serve as water storage and aquifer recharge. Surface flow maintains bay and estuarine system's nutrient and salinity balance. In the Keys, water supply and waste disposal are limiting factors
Limitations		Generally, minor limitations exist that can be corrected without excessive environmental degradation. Protection can be provided by on-site retention of storm water	Significant limitations exist due largely to flooding. Protection would require extensive drainage. Drainage should be authorized only after thorough, comprehensive land and water management plans have been established	Major environmental as well as structural limitations. Conventional development would require drainage to reduce flooding, thereby reducing wetland and groundwater supply	Severe environmental and structural limitations to development. Development would require drainage, removal of organic soils, and filling, destroying surface and underground water supply and quality and estuarine ecosystems. Storm surge hazard in coastal areas and Keys

ADAPTED FROM: South Florida Planning Council Study on Soils Types

regional 8 development constraints

STUART

LAKE OKEECHOBEE

PALM BEACH

SANIBEL ISLAND

NAPLES

FORT LAUDERDALE

MARCO ISLAND

TEN THOUSAND ISLANDS

GULF OF MEXICO

MIAMI

KEY BISCAYNE

ELLIOTT KEY

ATLANTIC OCEAN

FLORIDA BAY

FLORIDA KEYS

KEY WEST

north
mi.
0 5 10 20 30 40
km.
0 5 10 20 30 40 50 60

 1 None

 2 Minor

 3 Significant

 4 Major

 5 Severe

vegetation

Warm temperatures, high rainfalls, relatively flat terrain, and a high water table contribute to the composition of southern Florida's vegetation. In many parts of southern Florida, the native vegetation is the most conspicuous feature of the landscape. In a land of such low relief, vegetative types express physical features, which in turn mark drainage patterns (Craighead, 1971).

South of Lake Okeechobee, Florida is remarkably flat. Elevations at the northern end of the Everglades are about 17 feet above sea level, gradually sloping over a distance of approximately 100 miles to sea level at Florida Bay. The land generally falls less than two and one-half inches per mile. The topography, drainage, and vegetation are a highly integrated part of the ecosystem. Changes of inches in elevation can generate a completely different plant association, as illustrated by the hammock tree islands that dot the Everglades.

Alternating wet and dry climatic conditions coupled with slow drainage determine the duration of the hydroperiod (the length of the wet period when waterlogging exists at the surface of the upper soil layer). The Everglades saw grass marshes are dependent upon a long hydroperiod; the pine flatwoods prefer drier conditions although they can withstand an occasional long, wet period. The hydroperiod of the saw-palmetto prairies (the dry prairies) is generally shorter. The hydroperiod of swamps varies, ranging from long to continuous. The upland types of vegetation, such as the scrub high pine and the high hammock forests, are subjected to very short or no hydroperiods. Some plants can withstand short periods of flooding while others require long or continuous periods of inundation.

Plant communities naturally replace one another in orderly succession as the physical habitat is gradually changed by relatively transitory communities until the final or mature (climax) community evolves. In southern Florida aquatic plant communities, such as saw grass in freshwater or mangroves in salt and brackish water, produce organic matter which, when built up sufficiently, encourages the invasion of cypress or hardwoods. This type of succession depends upon many factors, including the hydroperiod, elevation of the water table, depth of organic matter, degree of sunshine, salinity, and availability of plant nutrients.

On higher, drier ground pines occur, giving way to hardwoods as underbrush encroaches. In the underbrush, shade-tolerant hardwood seedlings thrive, outcompeting the sun-loving pine seedlings. Pines often form a fire-subclimax community in which succession to the hardwood climax community is arrested. Frequent small ground fires thin the underbrush, and the resulting ash forms a good seedbed for the more fire-resistant pines.

Clearing and drainage alters plant succession. In many cases disturbed areas are invaded by exotic plants (plants not native to southern Florida). **Schinus terebinthifolius**, also called Brazilian Pepper tree or Florida Holly, the Cajeput tree (**Melaleuca quinquenervia**), and the Australian pine (**Casuarina** ssp.) have become established to the point where they are crowding out native vegetation.

John H. Davis, Jr., of the University of Florida, prepared the most comprehensive and definitive map of the southern Florida landscape in 1943. Figure 9 has been adapted from this work. Taylor Alexander of the University of Miami in a recent study of the vegetative changes concludes that "the major physiographic regions as established by Davis are still well defined and the overall appearance of the landscape has not been altered greatly as far as vegetation types are concerned, except where exotics have invaded" (Alexander, 1973).

The establishment of native vegetation in ornamental landscaping and in agricultural hedgerows can retard exotic plant invasions, help preserve indigenous plant species, and provide habitat for wildlife. Such self-sustaining areas require little or no maintenance and irrigation. Furthermore, native vegetation effectively integrates the urban and wilderness systems. Many landscape painters and photographers have captured the beauty of the southern Florida landscape; their works illustrate the vegetation available to enhance the built environment.

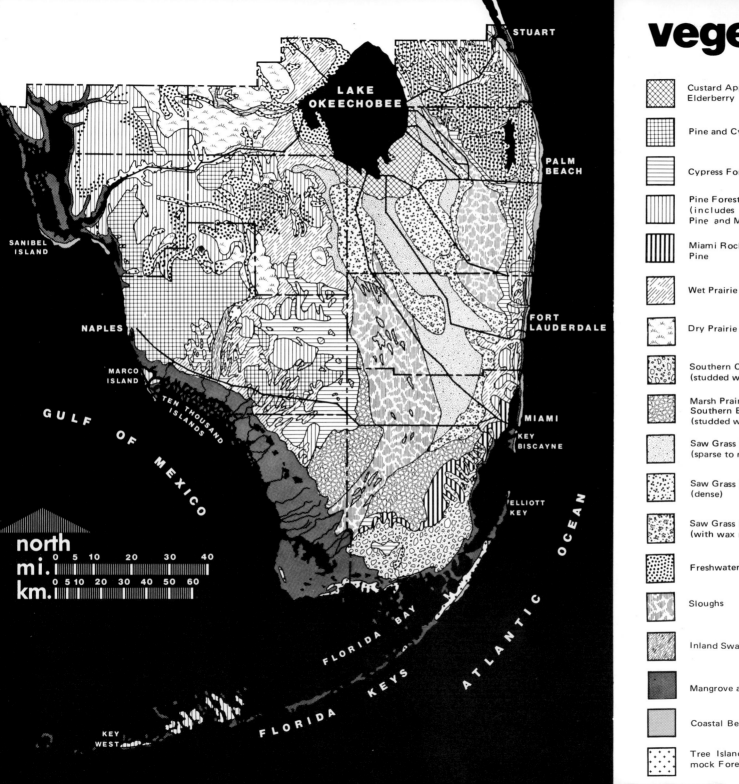

vegetation 9

 Custard Apple, Willow, and Elderberry

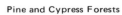 Pine and Cypress Forests

 Cypress Forests

 Pine Forests (includes Pine Flatwoods, High Pine and Miami Open Pine)

 Miami Rockland Pine

 Wet Prairie

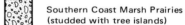 Dry Prairie or Saw Palmetto

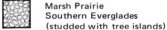 Southern Coast Marsh Prairies (studded with tree islands)

 Marsh Prairie Southern Everglades (studded with tree islands)

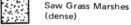 Saw Grass Marshes (sparse to medium)

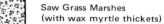 Saw Grass Marshes (dense)

 Saw Grass Marshes (with wax myrtle thickets)

 Freshwater Marshes

Sloughs

Inland Swamps

 Mangrove and Saltwater Marshes

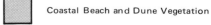 Coastal Beach and Dune Vegetation

 Tree Islands, Bay Tree, and Hammock Forests

climate

The subtropical climate of southern Florida is greatly influenced by its geographical location. Extending from approximately 27°N at Lake Okeechobee to 24°30'N at Key West, southern Florida is nearer to the equator than is any other part of the continental United States.

No point in the peninsula is more than 70 miles from the sea. The warm climate results primarily from three factors: (1) lack of prolonged cold outbreaks and the fact that cold fronts frequently stall near Lake Okeechobee, (2) many days of relatively intense solar radiation, and (3) proximity to the Gulf of Mexico and the Gulf Stream, which flows around the western tip of Cuba and passes a few miles east of the Atlantic coast. The climate of the region is essentially subtropical-marine. The prefix "sub" differentiates the climate from a true tropical climate since frosts, although rare, do occasionally occur.

The climate is characterized by balmy winters and long, warm summers punctuated by afternoon showers. During the period from May to October, southern Florida receives more rain than any other section of the United States. Ocean breezes, which are cool in the summer and

relatively warm in the winter, moderate the temperature. Along the east coast, water temperatures range from 72° in January to 80° in July. On the west coast, surface waters of the Gulf range from 70° in February to 84° in August (Bradley, 1972).

Precipitation, temperature, and wind directly influence the hydrology and water resources of the area. Climate together with geology have contributed to the shape of the land, the drainage characteristics, and ultimately the characteristics of plants, wildlife, and life-style of residents and visitors. Precipitation and wind data are detailed here since they can cause construction delays and influence the degree of air pollution.

Precipitation

Rainfall in southern Florida varies in annual amount and seasonal distribution from an average five-feet (60 inches) of rainfall per year on the mainland to almost 40 inches per year in the Keys (Appendix A). Rainfall patterns and quantities have a significant effect upon agriculture, tourism, water resource management, and land use planning. Figure 11 illustrates the monthly rainfall patterns and rates for southern Florida. Traditionally, about 70 percent of the rainfall each year falls between May and October. The remaining 30 percent falls between November and May, during which time evapotranspiration (loss of water vapor by plants to the atmosphere) almost always exceeds rainfall. This dry period each year usually depletes water stored in the Everglades Basin and the shallower aquifers. When rain does fall in the winter months, it is usually light and accompanied by overcast skies which, associated with the passage of fronts, may last for days. The beginning and end of the rainy season vary considerably from year to year, as do the time, location, and extent of precipitation.

Drought Conditions

The sufficiency of water supplies in southern Florida depends upon the intensity and frequency of droughts. A drought may be defined as a period in which rainfall has been so deficient as to hinder the growth of native vegetation and diminish water supplies. The effects of droughts may be aggravated by variations in temperature and wind speed. Above-normal temperatures, coupled with increased wind velocities, tend to increase the rate of transpiration from vegetation and evaporation from soil and water masses. Lower temperatures and decreased wind tend to decrease evaporation.

Winds

The prevailing winds over southern Florida are east and southeast. Inland, the average wind speeds can drop to half those measured along the coast. Figure 11 illustrates prevailing wind directions and speed. Appendix A lists the normal, means, and extremes of the climatic characteristics in southern Florida. High local winds of short duration occur occasionally in connection with thunderstorms in summer and with cold fronts moving across the state in other seasons. The air over the state often is unstable, a condition conducive to the development of cumulus clouds and thunderstorms. Sea breezes occur almost daily in the summer months, when other weather systems are relatively weak. These breezes are effective for several miles inland, even occasionally up to 30 miles from the coast.

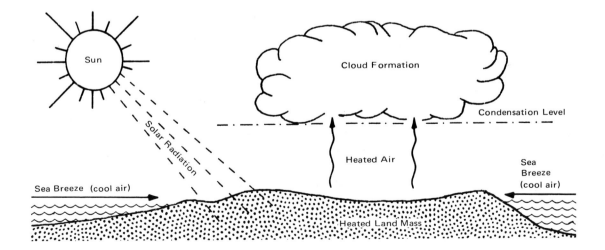

FIGURE 10:

SEA BREEZE: During the day as the peninsular land mass is heated by solar radiation, the air above it is also heated and begins to rise. Cooler humid air from the ocean moves in to replace the rising air and is heated in turn. As the air mass rises, it cools and is no longer able to hold as much moisture, so condensation (cloud formation) typically occurs. The clouds may become so saturated that precipitation (rainfall) begins. In the evening as the land mass loses heat rapidly to become cooler than the ocean surface, the situation is reversed, and a land breeze moves toward the sea.

Temperature

Mean annual temperatures range in the middle 70's in the southern Florida mainland, to nearly 78° in Key West (Appendix A). July and August temperatures are the warmest in all areas, and December and January temperatures are the coolest. Extreme heat waves, characteristic of northern continental locations, are rare, and temperatures of 100° are practically unknown. Temperatures are kept moderate by sea breezes and frequent afternoon or evening summer thunderstorms.

Other Climatic Characteristics

The climate of southern Florida is humid. Inland areas with greater temperature extremes enjoy slightly lower relative humidity, especially during hot weather.

Humidity varies little from place to place. Heavy fogs, which occur less than 10 days a year, are usually confined to the night and early morning hours in late fall, winter, and early spring. These fogs usually dissipate or thin soon after sunrise, and heavy daytime fog is seldom observed. Because the extensive wind patterns have crossed bodies of water, hot, drying winds seldom occur. Long hours of sunshine (about 66 percent of the daylight hours), comparatively longer winter days, high solar radiation, and the warming effect of the Gulf Stream all contribute to southern Florida's mild winters.

FIGURE 11: Monthly average rainfall (inches) patterns and prevailing wind direction and velocities. (Adapted from Gerrish, 1972b; Wood and Fernald, 1974.)

Wind speed (miles per hour)

← ● over 10

← ● 6 to 10

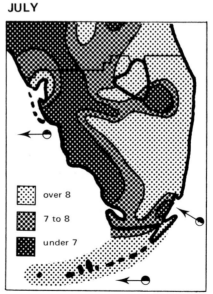

JANUARY

over 3
2 to 3
under 2

JULY

over 8
7 to 8
under 7

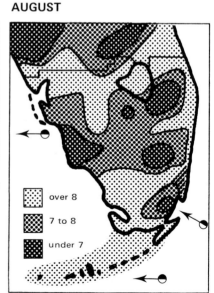

FEBRUARY

over 3
2 to 3
under 2

AUGUST

over 8
7 to 8
under 7

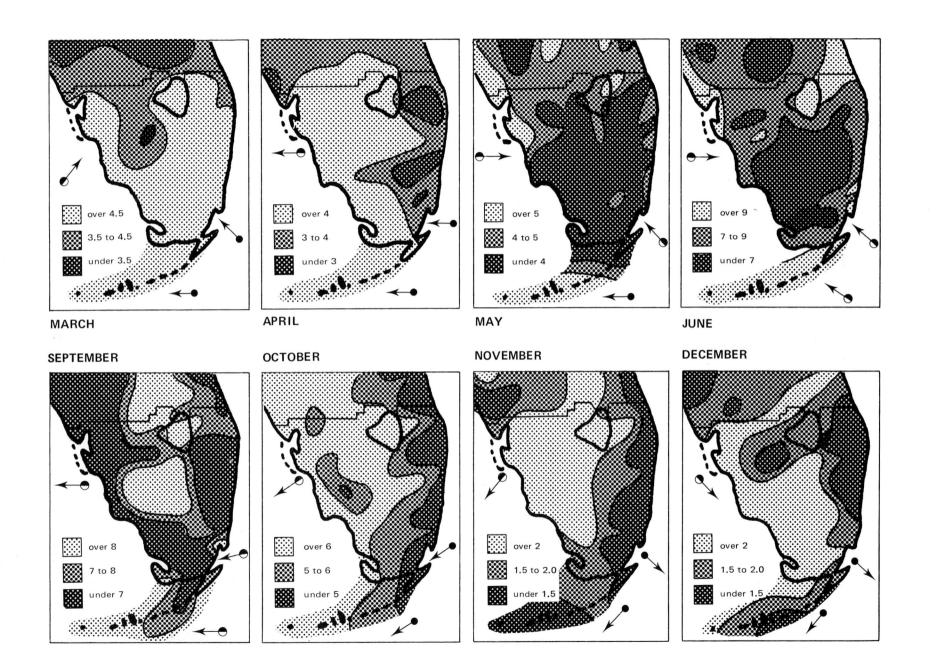

MARCH | APRIL | MAY | JUNE

over 4.5
3.5 to 4.5
under 3.5

over 4
3 to 4
under 3

over 5
4 to 5
under 4

over 9
7 to 9
under 7

SEPTEMBER | OCTOBER | NOVEMBER | DECEMBER

over 8
7 to 8
under 7

over 6
5 to 6
under 5

over 2
1.5 to 2.0
under 1.5

over 2
1.5 to 2.0
under 1.5

water resources

The prime source of water in southern Florida is rainfall. Precipitation falling upon the region averages five feet (60 inches) per year.

Hydrologic Cycle

The land-water-air interrelationship is referred to as the "hydrologic cycle." Simply illustrated, the cycle is the continuous process through which water moves from bodies of water (principally the ocean) to the atmosphere, falls to the land, and flows back to bodies of water (see Figure 12).

In southern Florida the ocean and extensive wetlands form the major source for most rain falling on the land. Air masses above the oceans accumulate the moisture that is evaporating from the sea. As this moisture-laden air passes over warm land surfaces, the air rises and the moisture condenses into rain-producing clouds. Moisture also enters the atmosphere via evaporation from bodies of water, from wet soils, and from transpiration by plants. In southern Florida the evapotranspiration loss has been estimated to consume about 75 percent to 95 percent of annual rainfall (U.S. Department of Interior, 1969, p. 18).

Rainfall percolates into the ground in proportion to the permeability of the surface upon which it falls. The remainder of the rain that is not absorbed forms temporary ponds or runs off into bodies of water.

Within the water cycle, there is an endless sequence of additive and subtractive factors, which over a long period of time must balance, both in time and quantity. This calculation is referred to as the water budget (i.e., total inflow of rainfall must equal total outflow of water). Evaporation and drainage from the land are continuous processes, variable in rate and amount throughout the region. Figure 17 is a diagram showing the theoretical distribution of average annual rainfall upon one square mile of Dade County. Although the amounts will vary throughout the region, this illustration, when compared with Figure 12, shows the interrelationship of the urban and natural communities within the total water cycle and water budget.

The hydrologic cycle of southern Florida is unique in its energy distribution and its high evapotranspiration rate. The interaction between land, air, and water is so closely interdependent that only parts of the system's matrix have been studied fully. The large percentage of rainfall lost to evaporation is due primarily to the intense heat of the subtropical climate and the large expanses of wetlands. The lush vegetation of southern Florida expedites transpiration and thus helps in the loss of large quantities of groundwater.

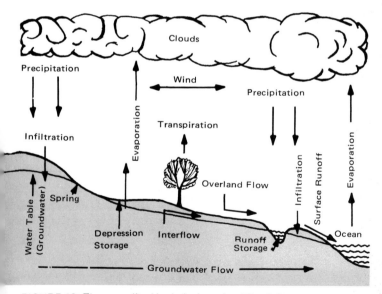

FIGURE 12: The generalized hydrologic cycle of South Florida.

34

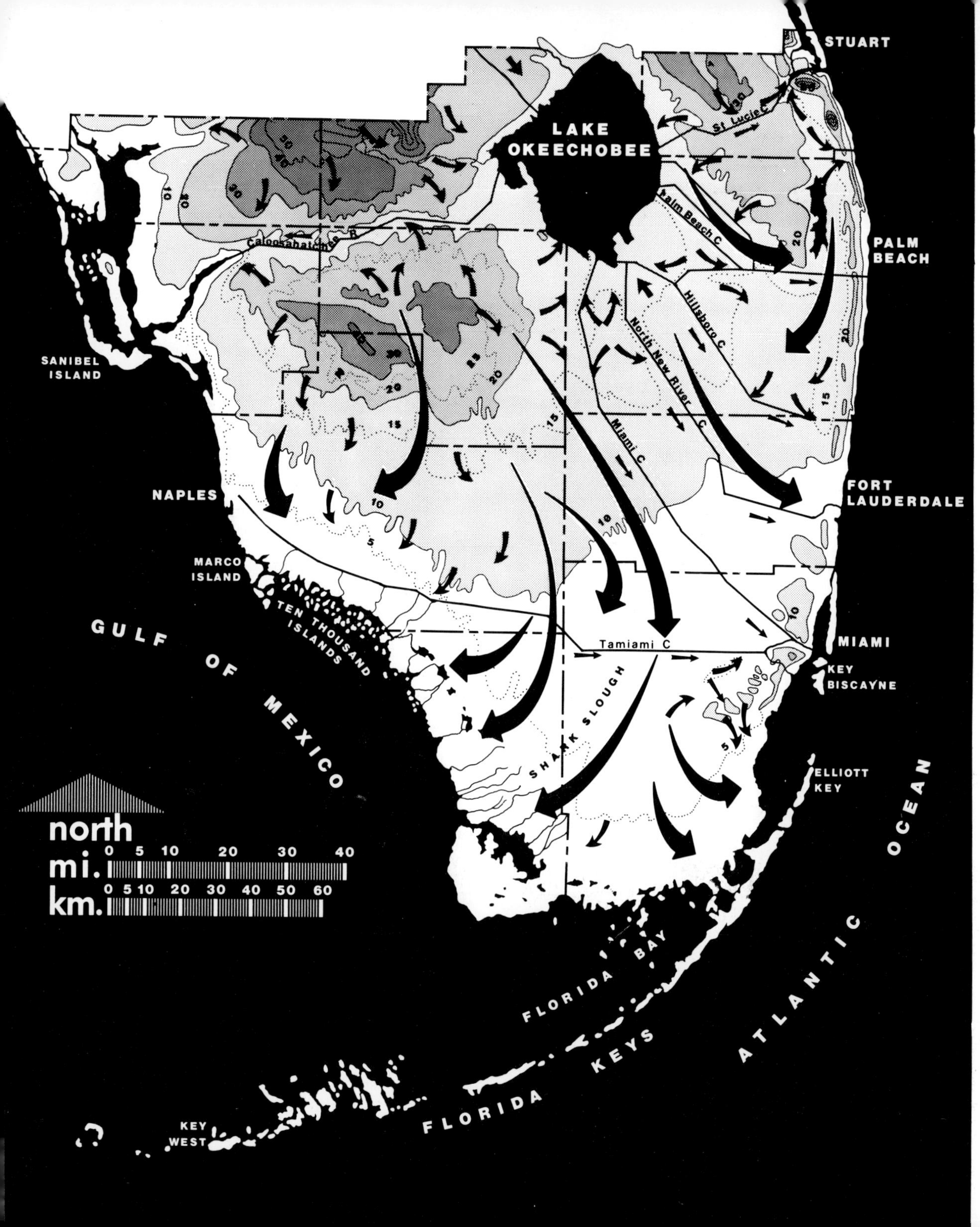

STUART

LAKE
OKEECHOBEE

PALM
BEACH

13
topography
water flow

SANIBEL
ISLAND

Caloosahatchee R.

Palm Beach C.

Hillsboro C.

North New River C.

Miami C.

FORT
LAUDERDALE

NAPLES

MARCO
ISLAND

TEN THOUSAND
ISLANDS

Tamiami C.

MIAMI

KEY
BISCAYNE

GULF OF MEXICO

SHARK SLOUGH

ELLIOTT
KEY

north
mi.
km.

| 0 | 5 | 10 | 20 | 30 | 40 |
| 0 | 5 | 10 | 20 | 30 | 40 | 50 | 60 |

FLORIDA BAY

FLORIDA KEYS

ATLANTIC OCEAN

KEY
WEST

Elevation in feet

 40' +

 30' – 40'

 20' – 30'

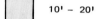 10' – 20'

 0' – 10'

⌢ 10 foot interval

⋯ 5 foot interval

➜ Direction of water flow

SOURCE

Adapted from Florida Geological Survey, Bulletin 25, Figure 26 (Topographic and Drainage Map of Southern Florida), J. H. Davis, Jr., "Natural Features of Southern Florida" (1943).

35

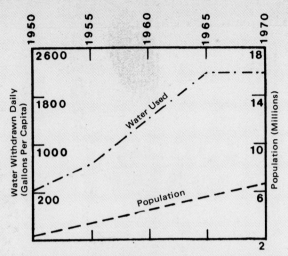

FIGURE 14: Trends in population and withdrawals of water in Florida, 1950 to 1970 (adapted from U.S. Geological Survey, 1973, fig. 6, p. 29).

FIGURE 15: Location of major well fields.

consump-tion

Water is vital to all life support systems. Demands from an ever-increasing per capita consumption rate, coupled with somewhat wasteful water management practices, may soon exceed the system's supply (Hartwell, 1973).

Figure 16 shows the geographic extent of the principal aquifers of southern Florida. The Biscayne Aquifer is considered to be one of the mostly highly productive aquifers in the world. Extending roughly from southern Palm Beach County to southern Dade County, this valuable natural resource provides the drinking water for the three counties in its domain as well as a major supplement to the island communities of Monroe County. The other aquifers illustrated here are considered to be less productive and of a lower quality.

Figure 15 shows the location of major well fields in southern Florida. As one would expect, they correlate well with centers of population. Daily water consumption figures vary greatly among urban areas of the United States. The 1970 national average of water consumption equaled 166 gallons per capita per day (gpcd), and the average for the entire state of Florida was 163 gpcd. Southern Florida averaged 196 gpcd.

Figure 14 shows that water consumption in Florida is increasing more rapidly than the number of people. Among the more significant causes for the higher water consumption rate in southern Florida is its subtropical climate, characterized by a hot, humid rainy season, and a cooler dry season. Water demand is high for cooling and sanitation purposes during the hot, rainy season and for lawn irrigation during the dry season. Furthermore, the majority of southern Florida communities are on the seacoast where a great influx of tourists from the north increase consumption during the dry winter season.

Faced with a growing demand that may exceed the available supply of water, government and industry have been reexamining the way in which the limited water supply is managed and how it is used. Reduced water supply through extensive drainage, pollution of water, and the increasing per capita demand can affect the quality of life for the citizens of southern Florida.

Water Flow

The flow of water in the region is affected by natural topography, and more recently by canals and levees within the Central and Southern Florida Flood Control District (FCD).

Figure 13, "Topography and Water Flow" illustrates the general water movement and topographic relief of southern Florida. Lake Okeechobee receives the surface flow from some 4,700 square miles of land and water areas of the Kissimmee River basin lying to its west and north. Divides are poorly defined in many places, and the direction of surface drainage often is determined by the distribution of recent rainfall.

The water cycle of southern Florida functions in two ways: groundwater recharge and, to a lesser extent, surface water flow. Historically, the only significant surface flow to tidewater occurred at the southern tip of Florida, now within the Everglades National Park. Collier County receives no water from Lake Okeechobee, and, in fact, rainfall falling on the northeastern section of the county flows into Conservation Area 3. In Monroe County, the Florida Keys depend entirely on rainfall; the mangrove swamps along Florida Bay and the Gulf of Mexico are largely watered by tides, summer rains, and, during the rainy season, overflow from the freshwater swamps behind the Buttonwood Levee (Craighead, 1971).

The coastal ridge running parallel to the coast from Palm Beach through Dade County forms the eastern boundary of the Everglades waterway. Water flow in the Everglades curves to the southwest, away from the coastal ridge, and its course is outlined by the elongation of the tree island hammocks that characterize the Shark River Slough flowing through west Dade County.

Water tables throughout southern Flor- ida are at or near the surface, and excavations quickly fill with groundwater. Many factors determine the rate and amount of surface flow within the region, including topographic relief, level of the water table, absorption of soils, vegetation cover, etc., and therefore on-site inspection and more detailed site information is necessary for the analysis of a particular area within the region.

Water Management

The Everglades Drainage District (now the Central and South Florida Flood Control District) was created by the 1905 Florida Legislature to expedite the "reclamation" of southern Florida. The hydrology of southern Florida was significantly affected before 1945, when land drainage and reclamation were the principal objectives of the District. Estimates indicate that water levels were lowered considerably, perhaps as much as five or six feet as a result of uncontrolled drainage.

Following the extensive flooding of southern Florida and the east coast in 1947 and 1948, the United States Army Corps of Engineers received authorization to proceed with an extensive system of canals and levees. The FCD was established in 1949 with its primary concern being flood control, agricultural irrigation, and providing water for growing urban needs. The period from 1948 to 1970 was essentially one of water quantity manage-

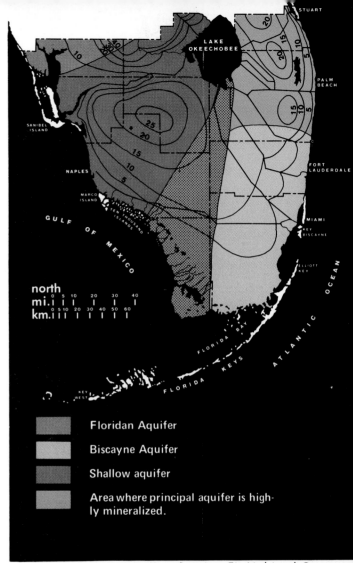

FIGURE 16: Generalized aquifers of southern Florida (above). Section of the Biscayne Aquifer in Dade County (below). (Hartwell, 1973).

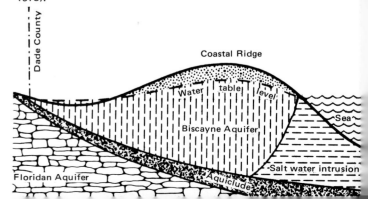

ment as well as flood control. Figure 18 shows the FCD plan under which the policies were implemented.

In the early 1970s, FCD reviewed its policies and moved toward becoming a water conservation agency, with emphasis upon water quality as well as quantity. The effectiveness of the present water control system of southern Florida is questioned by many. In 1972, the Florida Legislature passed the Water Resources Act and the Environmental Land Management Act, two landmark laws that bring coordinated planning and management of these natural resources together under appropriate state and regional agencies. The drastic effects of water control practices and policies demand that close scrutiny be given to all future projects, with the maintenance of the viable, functioning ecosystem for urban communities and nature being given the highest priority.

flood and storm

The flood potential of any given area is determined by the presence of physical conditions such as water levels, soil saturation, ground elevation, topographic relief, and the permeability of the surface. Southern Florida's relatively low topographic relief, which varies only from a few feet in elevation in most places to slightly over 50 feet above sea level in others, offers little protection from major storm floods. Heavy rainfall causes periodic flooding in the interior; storm surge flooding affects the coastal zones. Early warning systems by the National Hurricane Center have greatly reduced the loss of human life, although extensive property damage still occurs in flood-prone areas. Many Floridians have accepted this ever-present threat as a way of life; but, since southern Florida is becoming a major population center, it is necessary that future development patterns recognize this threat to life and property and consider it in designs for the total environment.

Coastal Zones

Coastal flooding is caused by storm surges from the ocean, and may be intensified by heavy rainfall. The extent of flooding in the coastal zone is greatly influenced by the elevation of the coastal

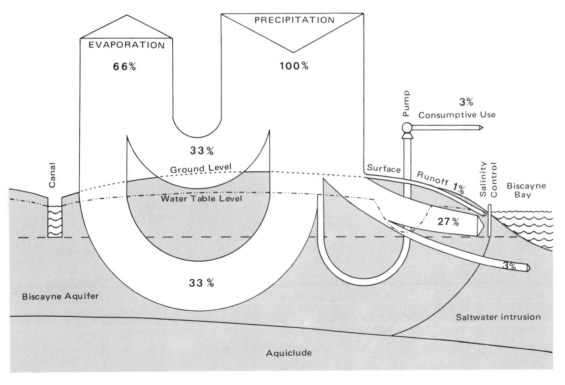

FIGURE 17: Diagram showing the theoretical distribution of average annual rainfall on one square mile of coastal Dade County (adapted from Meyer, 1971).

area and by the form of the shoreline. Where there is no coastal ridge the rate of fall-off of the storm tide is estimated to be about one foot per mile inland. Thus an uninhibited 15-foot storm tide could theoretically extend up to 15 miles inland. Coastal vegetation can moderate this surge by increased surface friction and absorption of wave energy. If, however, a coastal ridge with an elevation of ten feet or more parallels the shoreline, at least five miles inland, it is unlikely that any storm tide would wash over the ridge.

Interior Zones

When the inland areas are saturated by heavy rainfall, the Central and South Florida Flood Control District (FCD) system of canals and levees is capable of draining one-half inch of water daily. The canal and levee system is shown in Figure 18. When an overall rainfall of two inches or more is expected, the Miami Weather Center notifies the FCD, which then begins to discharge as much water from the canals as possible prior to the expected rainfall. This action increases the capacity of the system to carry off the anticipated rainfall. Although this major drainage system is considered to be vital to the protection of life and property from flood damage, many natural scientists, government agencies, and to some extent the FCD itself, are questioning the impact of the system on the region's ecosystem. The flood control system has lowered water tables,

allowed saltwater intrusion, and increased the probability of storm surge being conducted inland via the canals. Unfortunately, the exact correlations between all the variables involved in determining the timing and severity of interior flooding have not yet been computed by any agency.

Figure 21 illustrates the coastal and inland flooding potential to southern Florida under severest conditions. Along the Atlantic coast, the coastal ridge protects inland areas from storm surge flooding, but flooding can occur when heavy precipitation combined with major surge conditions raises the water level, breaching the numerous channels through the ridge. The western coastline and offshore islands of the Gulf of Mexico are less protected. The severity of flooding is also dependent upon the angle at which the storm strikes the shoreline. An angle of approximately 90° (almost perpendicular to the coast) will cause maximum damage. Shoreline configuration further modifies storm surge effects. Impact is not as severe where the shoreline is convex, as in the Naples area, but can be intensified by a concave form such as Biscayne Bay. The effects of storm surge are maximized in bays where a spit or bar prevents the water from returning quickly to sea. (See Figure 20.)

Hurricanes

Thirty-seven hurricanes have struck

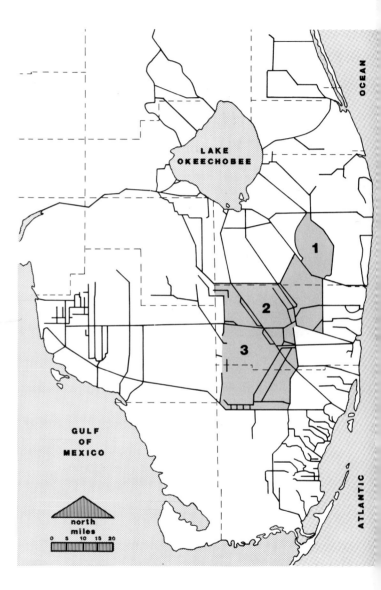

FIGURE 18: The water control works of the Central and Southern Florida Flood Control District.

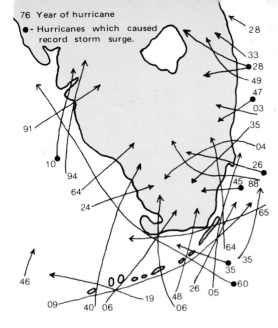

76 Year of hurricane
●– Hurricanes which caused record storm surge.

FIGURE 19: Hurricanes in southern Florida, 1888 to 1968. (Adapted from Wood and Fernald, 1974; Simpson, Frank, and Carrodus, 1969.)

FIGURE 20: Effects of coastal land forms on storm surge impact.

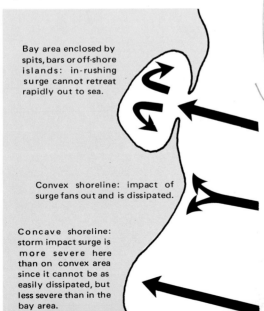

Bay area enclosed by spits, bars or off-shore islands: in-rushing surge cannot retreat rapidly out to sea.

Convex shoreline: impact of surge fans out and is dissipated.

Concave shoreline: storm impact surge is more severe here than on convex area since it cannot be as easily dissipated, but less severe than in the bay area.

southern Florida since 1885 (see Figure 19). According to the National Weather Service, hurricanes occur every six to eight years in this region. The "hurricane season" extends from June to November, with hurricanes occurring most frequently during September and October.

Hurricane Donna, which crossed the Keys and then moved northeastward across the state from near Fort Myers to near Daytona Beach on September 9-10, 1960, is thought to be the most destructive storm ever experienced in Florida. This storm caused an estimated $300 million in damage. Only 13 fatalities resulted from this storm, indicating the great value of the modern storm warning service now available in hurricane threatened areas. In 1964, hurricanes Cleo, Dora, and Isbell caused the greatest damage for any one year—$350 million.

Federal Flood Insurance Program

The Flood Insurance Program will have a significant impact on the form and location of future development in southern Florida.

All ten counties and many communities in the study area are included in HUD's list of communities having flood hazard areas. A special flood hazard area is defined as an area which has a one percent annual chance of flooding (or, is subject to the hundred-year flood).

The Flood Disaster Protection Act of 1973 requires HUD to notify communities that they have been designated as flood prone. The community must then either make prompt application for participation in the flood insurance program or must satisfy the secretary of HUD that the area is no longer flood prone. Participation in the program is required by July 1, 1975, or the community will be denied both federally related financing and most mortgage money.

Individuals and businesses located in identified areas of special flood hazard are required to purchase flood insurance as a prerequisite for receiving any type of federally insured or federally regulated financial assistance for acquisition or construction purposes. Effective July 1, 1975, such assistance to individuals and businesses will be predicated on the adoption of effective land use and land management controls by the community.

Federally subsidized insurance for flood hazard is authorized only within communities where future development is controlled through adequate flood plain management. Management may include a comprehensive program of corrective and preventative measures for reducing flood damage, such as land use controls, emergency preparedness plans, and flood control works. Participating communities may be suspended from the program for failure to adopt or to enforce land use regulations. (National Flood Insurers Association, 1974.)

flood hazard 21

STUART

LAKE OKEECHOBEE

PALM BEACH

SANIBEL ISLAND

NAPLES

MARCO ISLAND

TEN THOUSAND ISLANDS

GULF OF MEXICO

FORT LAUDERDALE

MIAMI

KEY BISCAYNE

ELLIOTT KEY

ATLANTIC OCEAN

FLORIDA BAY

KEY WEST

FLORIDA KEYS

north
mi.
0 5 10   20   30   40

km.
0 5 10   20   30   40   50   60

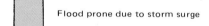 Flood prone due to storm surge

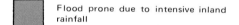 Flood prone due to intensive inland rainfall

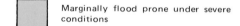 Marginally flood prone under severe conditions

 Area not flood-prone

Direction of storm surge

General direction of overland flow

SOURCE

Interview with Mr. John Hope, National Hurricane Center, Miami, Florida

41

wildlife

Southern Florida's wild fauna is as tuned to the season as the flora. This relationship was poetically described by ecologist Arthur Marshall (1972): "Heavy tropical rains come to south Florida in the summer and fall throughout the Everglades waterway. The water rises out of its shallow, scattered depressions—lakes, rivers, ponds, and sloughs. It sheets over the southern Everglades marshes in the form of a very broad river which historically was seven or eight feet deep at summer flood.

"Regenerative processes bloom with the rising spread of the water under the warm summer sun. Plant germination and growth flourish. Many aquatic organisms—insects, crayfishes, killifishes, reptiles—engage in an orgy or reproduction and in a few weeks their progeny can be seen in all reaches of the summer river.

"After the rains let up, the sheet water recedes into the deeper ponds and sloughs and concentrates the summer's production of small organisms, making them available in essential densities to the waiting large predators. This phenomenon of flood-bloom-recession-concentration is a marvel of synchronization—for the summer's organic products are thus served up to the flocks of colonial birds who are then fledging their young and to the young and adults of many marine fishes which invade the brackish and fresh waters of the lower Everglades to forage there."

The later winter dry-downs concentrate fish for anglers also. The late winter of 1971 afforded the best Everglades fishing in years; however, the stress resulting from such extreme drought conditions led to decreases in fish and land animal populations in subsequent years. Prolonged or nonseasonal flooding also has adverse effects—organic ooze builds up. In flooded areas, vegetation patterns change, and birds and deer cannot forage successfully.

Today only a remnant of southern Florida's native wildlife remains since one-half of the wetlands have been drained and the other half altered to varying degrees in the amount and periodicity of flooding. Animal species which are becoming rare in southern Florida are designated by asterisks in Table 5. They include species of birds and mammals, as well as reptiles.

In general, what man must do to protect wildlife is protect the habitat. The complete domestication of wilderness areas renders them fit only for man, domesticated animals, and those wild animals which can adapt to urbanization—lizards. raccoons, opossum, squirrels, and rats. Many animals can survive in areas which are carefully utilized by man. Hedgerows and moderately sized sloughs and woodlots will provide habitat for rabbits, song and game birds, turtles, snakes, foxes, and waterfowl. Slightly larger areas such as greenbelts will support deer, wild pigs, and bobcats. At the other end of the spectrum are the now rare Florida panther and black bear which have retreated to the deepest thickets of the Big Cypress Swamp.

Estuarine habitats are important also. At least 65 percent of all marine organisms spend some portion of their life cycle in the brackish waters of estuaries and the fringing mangrove swamp. Sea turtles require uncluttered, relatively unlit beaches for nesting and hatching of their young. The endangered manatee (sea cow) probably will survive if access to warm inland waters is maintained and the animals are protected from injury or molestation by boaters.

Environmental research continues to provide new insights into the ecology of wildlife. Its future depends largely upon man's willingness to make use of this knowledge and to value wildlife for its own sake.

TABLE 5

SELECTED LIST OF ANIMAL SPECIES OCCURRING IN SOUTHERN FLORIDA

SPECIES	DISTRIBUTION	PREFERRED HABITAT	STATUS
Mammals			
Black bear*	Sporadic, statewide	Inaccessible areas	Rare
Bobcat	Statewide	Swamps, flatwoods, hammocks	Formerly common, population on decline
Eastern cottontail	Uplands east and west coast	Brushy areas	Common
Everglades mink*	Big Cypress	Aquatic habitats	Rare
Florida panther*	Sporadic statewide, mainly Everglades	Inaccessible areas	Rare
Gray fox	Statewide	Flatwoods, scrub	Declining
Gray squirrel	Statewide	Mature forest stands, urban areas	Common but declining
Manatee (sea cow)*	Everglades National Park, Biscayne Bay and tributaries	Marine and estuarine areas	Rare
Marsh rabbit	Statewide	Swamps, scrub, flat areas	Common
Opossum	Statewide	Swamps, scrub, flatwoods, urban areas	Common
Otter	Statewide	Aquatic habitats	Declining
Raccoon	Statewide	Flatwoods, swamps, scrub, urban areas	Common
White-tailed deer	Statewide	Flatwoods, swamps, scrub	Common and increasing in drained areas
Wild pig (feral)	Statewide	Swamps, flatwoods, scrub	Must be stocked
Round-tailed muskrat	Statewide	Aquatic habitats	Common but declining

SPECIES	DISTRIBUTION	PREFERRED HABITAT	STATUS
Birds			
Bald eagle*	Big Cypress, Everglades	Wetlands, estuaries	Rare
Brown pelican*	Statewide	Estuarine	Rare
Cape Sable sparrow*	Cape Sable area	Spartina marshes	Rare
Mourning dove	Statewide	Ubiquitous, farmlands, urban areas	Common
Ducks	Statewide	Aquatic habitats	Fluctuating
Everglade kite*	Everglades, Loxahatchee Wildlife Refuge, FCD Conservation Areas	Aquatic habitats	Rare
Great white heron*	Florida Bay, Keys	Marine and estuarine areas	Rare
Limpkin	Statewide	Wooded swamps	Declining
Osprey*	Marine and freshwater areas	Estuaries	Rare
Peregrine falcon*	Migratory	Coastal wetlands	Rare
Quail	Statewide	Farmlands, flatwoods	Fluctuating
Red-cockaded woodpecker*	Statewide	Pinelands	Rare
Roseate spoonbill*	Florida Bay	Aquatic areas	Rare
Sandhill crane*	Scattered	Freshwater marshes, flatwoods	Rare
Short-tailed hawk*	Scattered	Hardwoods, scrub	Rare
Wild turkey	Statewide	Flatwoods	Increasing
Wood stork*	Big Cypress	Swamps	Rare

Reptiles

SPECIES	DISTRIBUTION	PREFERRED HABITAT	STATUS
American alligator	Everglades, Big Cypress	Fresh and brackish water areas	Formerly rare, increasing
American crocodile*	Northern Keys, southern Florida coastline	Remote estuarine and coastal regions	Rare
Green turtle*	Atlantic Ocean, Gulf of Mexico	Marine; nests on coastal beaches	Rare

*In Table 5 species are designated rare and endangered on a continentwide basis by the U.S. Department of the Interior. Southern Florida is one of the last remaining refuges for species that are in many cases extinct in other areas of North America where they once flourished in great numbers. These species may still be numerous in southern Florida either as resident species or seasonally when migrants gather here from the north.

SOURCE: Rodgers and Crowder, 1974; Owre, 1975.

ECOLOGICAL REGIONS

The ecological regions delineated on the map show remarkable diversity for a terrain that varies only 25 feet in elevation from the coastal ridge to the mud flats of Florida Bay. The southern Florida region is saucer-shaped, rimmed by the sandy flatlands and coastal ridge, with the low-lying Everglades and Big Cypress occupying the bowl. The saucer is tipped slightly to the south, and surface water slowly flows over the southern rim into Florida Bay and the Gulf of Mexico.

Minor differences in elevation of a foot or two, accompanied by variation in water depth or flow, can change the character of the vegetation completely. In the flatlands, for example, a sinkhole a few feet across can hold a miniature wetland community of willows or cattails in the midst of the drier pines. This response to small differences creates distinct topographic-ecological regions expressed in typical vegetation-soil-hydrology associations for each region.

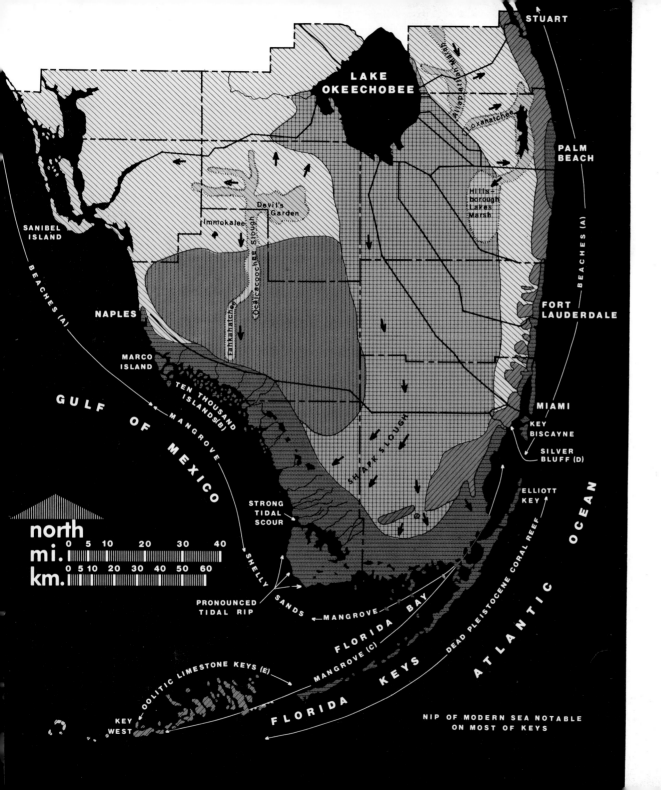

22
ecological regions

COASTAL MARSHES AND MANGROVE SWAMPS: Very low, wet areas of marly to mucky soils thinly overlying rock.

BIG CYPRESS SWAMP: Flat, poorly drained, with thin marly or mucky and sandy soils and bare areas of solution-riddled limestone.

SANDY FLATLANDS: Low-lying, defectively drained lands, generally flat although parts on higher terraces are gently rolling.

ATLANTIC COASTAL RIDGE: Hardly discernible as a ridge. Generally underlain by rock, in places exposed at the surface. Cut through in places, especially in the southern part, by old tidal and natural drainage channels.

EVERGLADES: Organic soils of peat and muck with widespread deposits of calcareous marl, almost completely flat surface, generally wet throughout the year except where drained.

A   **BEACHES:** Long, sandy, and shelly with spits and bay bars.

B   **TEN THOUSAND ISLANDS:** Many of these are drowned Pleistocene sand dunes.

C   **INTRICATE MANGROVE SHORELINE:** Found on both bay and ocean sides of the Keys as well as along the mainland.

D   **SILVER BLUFF, LOW SEA CLIFF:** This shows tip of Pleistocene sea at 8' and 5' above present mean sea level.

E   **HIGHEST KEYS:** These are forested with Caribbean pines.

SOURCES

U.S. Geological Survey, Water-Supply Paper 1255, Plate 12 (Topographic-Ecologic Map of Southern Florida), "Water Resources of Southeastern Florida" by Garald G. Parker, G. E. Ferguson, S. K. Love, et al. U.S. Government Printing Office, Washington, D.C. (1955).

The Sandy Flatlands

Rainfall here either percolates rapidly into the sandy soil or is stored in the circular ponds that dot the region. During the rainy season, runoff is slow and flooding is common in areas of low permeability because the surface is flat. Most of the drainage occurs through the Devil's Garden-Okaloacoochee Slough-Fakahatchee system in the western flatlands and via Allapattah Marsh and Hongry Land-Loxahatchee system in the east. These drainageways, where water is found most of the year, contain a myriad of species typical of freshwater swamps. Cypress stands, water oaks, true hollies, willows, and orchids of many kinds provide a striking contrast to the open slash-pine forest of the surrounding flatlands. On high fossil dunes occasional scrub forests occur, containing sand pine and several oak species, but little ground cover.

The Coastal Ridge

Along the Atlantic coast, a higher strip, hardly discernible as a ridge, separates the eastern flatlands and Everglades from the beaches, dunes, and coastal marsh of the coast. North of Miami the ridge is overlain by sands, but toward the south the underlying rock emerges frequently at the surface. The ridge disappears south of Homestead where the Everglades drain into Florida Bay, emerging again in the lower Keys.

Open forests of pine and palmetto dominate the ridge. Along the eastern margin saw palmetto prairies form a transitional community as the ridge dips into the Everglades. The pine woods are interspersed with hammocks of various types where local conditions and absence of fire allow soil formation to occur. The pinelands, on the other hand, require periodic fire to prevent the hardwoods of the hammocks from encroaching.

The Everglades

The shallow basin of the Everglades overlies relatively impermeable marl and limestone formations, at least in the north and central portions. Rainfall cannot percolate downward, but instead slowly flows over the surface toward the south. This water, stored at the surface, eventually reaches areas of permeable sand or limestone where aquifer recharge can occur. Much of the natural drainage has been altered to supply water for storage in the Conservation Areas.

The peaty organic soils covering the marl and limestone are formed of partly decomposed plant remains, mostly of the dominant plant, saw grass. The treeless expanse of the saw grass glades, the famous "River of Grass," is interspersed with tree islands, or "heads," particularly in the southern portion. These slightly elevated islands, which are aligned with the direction of water flow, support a distinctively different vegetation than the watery glades. Willow heads, bay heads, and cypress domes are named for their dominant tree species.

Big Cypress Swamp

A low, limestone ridge separates the slightly lower Everglades from the Big Cypress; on the north and west are the flatlands. In contrast to the sandy or peaty soils of these adjoining regions, the Big Cypress has large areas where bare, pitted limestone forms the surface. Thin layers of marl may lie in old drainageways or depressions in the rock. Most of the swamp is flooded during the rainy season, and the surface water flows almost imperceptibly toward the coast. Here in the Ten Thousand Islands the drainage becomes more positive through the many intricate creeks and rivers of the mangrove swamp.

The floor of the Big Cypress is highly dissected by solution holes and swales. In higher areas with better aerated soils the dominant palmettos, pines, and grasses give way in places to hammock growth. Cypress, bay, willow, and marsh vegetation are found in the lower wetlands.

Mangrove and Coastal Marsh

Mangroves develop on shorelines of lower wave energy, in contrast to beach development. They are land builders, and the swamps in southwestern Florida rest on about 12 feet of mangrove peat deposits mixed with sand. The slowly rising sea would otherwise have inundated these coastal areas. Mangroves also protect the shore from erosion by storm action. One of the best developed mangrove forests in the world cloaks the Ten Thousand Islands.

Inland from the mangrove fringe lie the coastal marshes, a transitional strip where marl soils predominate. The salt-tolerant marsh association near the sea, where occasional tidal inundation occurs, gives way to freshwater marsh species further upland.

The Florida Keys

The upper Keys, from Soldier Key to Bahia Honda, are ancient coral formations, lying parallel to the edge of the Floridan Plateau. From Big Pine Key to Key West the lower Keys are oriented at right angles to the upper Keys. This reflects a difference in origin, the lower Keys being an extension of the same oolitic formation as the coastal ridge. The channels between the lower islands were formed by tidal scour, and in orientation and spacing are remarkably like the transverse glades that cross the ridge on the mainland.

Keys vegetation is highly variable. Mangroves and hammocks are found throughout the Keys, but from Big Pine Key southward the larger islands are also forested with the slash pine of the coastal ridge. The climate is more tropical than that of the mainland, and some Caribbean species that flourish here are found nowhere else in the continental United States (Parker, Ferguson, and Love, 1955; Long and Lakela, 1971.)

land use

The use of the land in southern Florida began with both courage and ignorance—courage to inhabit a land that appeared to favor fish and wildlife, and ignorance that led to the destruction of parts of the richness of the region by indiscriminate clearing, draining, and filling.

Less than a century ago, Florida began its march into the twentieth century to become one of the fastest growing populations in the nation. An expanding transportation system together with the availability of automobiles fostered the great Florida land boom of the 1920s. In 1975, the 10 counties of the southern Florida region are the home of over 3,000,000 people and the playground of almost 10 million tourists annually. The total land area of the region is 12,257 square miles.

Of the 12,257 square miles included in the region, only approximately 3,300 square miles are in community development; 2,824 square miles are managed conservation areas and parks; 4,969 square miles are in agriculture land. The remaining 1,164 square miles are considered undeveloped. It is in this remaining yet undeveloped acreage and the inefficient use of the existing urban land areas that future policies must recognize the interrelationship of man and nature.

The map on the opposite page shows the region's major urban concentration to be in the three "Gold Coast" counties: Dade, Broward, and Palm Beach. In Collier, Monroe, Lee, and Charlotte counties development has been more clustered.

The urban region is a fact of life. The vast majority of Floridians will probably continue to live in, what is a relatively small area of land, not in cities as we have known them, but in snakelike megalopolises tied together by roads and markets. The issue is not whether there will be urban regions in the future. The issue is what form and what level of environmental quality these urban centers will reach. (Rockefeller Brothers Fund, 1973.)

STUART

LAKE
OKEECHOBEE

PALM
BEACH

FORT
MYERS

1

SANIBEL
ISLAND

IMMOKALEE

2

CONSERVATION AREAS

FORT
LAUDERDALE

NAPLES

BIG
CYPRESS
PURCHASE
AREA

3

MARCO
ISLAND

TEN THOUSAND ISLANDS

MIAMI

KEY
BISCAYNE

GULF OF MEXICO

EVERGLADES
NATIONAL
PARK

ELLIOTT
KEY

north
mi. 0 5 10 20 30 40
km. 0 5 10 20 30 40 50 60

FLORIDA BAY

FLORIDA KEYS

ATLANTIC OCEAN

KEY
WEST

23
land use

Big Cypress Preserve
Purchase Area

Everglades National Park

State Park

Wildlife Management Area

Wildlife Refuge

Indian Reservation

Indian Reservation

Urban

Agriculture

SOURCES:

South Florida Regional Planning Council,
Florida Department of Transportation.

demography

Any viable plan for the future development of a given area must have at its core a sound projection of the future population characteristics of the area. A series of population projections were calculated, using different methodologies. The projections are shown first for the ten-county region as a whole, and then each county is analyzed on an individual basis.

The sources for these projections are:
SERIES I: The Office of Business Economics, Bureau of Census, U.S. Department of Commerce, as disaggregates by the Atlanta office of the Environmental Protection Agency.
SERIES II: Jerome P. Pickard, Research Director, Urban Land Institute, Washington, D.C.
SERIES III: An extrapolation using the 1950-1970 growth rate and projecting it to the year 2000.
SERIES IV: The Florida Social Science Advisory Committee.
SERIES V: Center for Urban and Regional Studies, University of Miami.

As illustrated by Figure 24, the projections of Series I and IV are almost identical even though they were calculated by two unrelated groups working independently. Series II, which was provided by the Research Director of the Urban Land Institute, is less conservative. While Series I and IV project an annual growth rate of roughly two percent, the Urban Land Institute projects a growth rate of just over three percent. Series III is a simple extrapolation of the growth that occurred in southern Florida from 1950 to 1970; it projects a continuing annual rate of six percent which would give Florida a population of over 12 million people by the turn of the century. For purposes of this study, a fifth series has been projected by the Center for Urban and Regional Studies as a sort of midline composite, and the specific figures appear in Table 6. (Nicholas, 1975.)

There is general agreement among demographers that the percent growth rate that has characterized the Florida scene for over 20 years will eventually begin to decline. The only point of uncertainty is when this decline will take place.

It is not anticipated that Florida's population will decline or that growth will stop but that the rate of growth will begin to diminish as southern Florida approaches its carrying capacity. Carrying capacity depends upon economic as well as physical factors (e.g., the availability of investment capital, freshwater, raw materials, energy). The major regulator will be economic; highly developed and heavily populated areas consume enormous quantities of energy, and the cost of living in such areas can be expected to rise with the rising cost of energy.

Individual analyses of the ten counties of southern Florida follow:

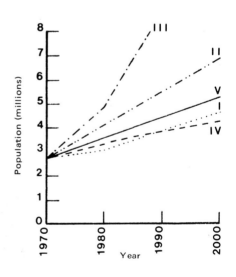

FIGURE 24: Population projections for the ten county southern Florida region (adapted from: Nicholas, 1973).

Broward County

In the twenty years from 1950 to 1970, Broward County was the fastest-growing county in the United States. The county grew at a rate of over 10 percent a year compared to slightly over six percent for the region as a whole. The majority of the growth took place in southeastern Broward County for the first 15 years and then spread to the west. During this period the city of Hollywood grew from about 35,000 persons in 1960 to 107,000 in 1970 (just over one-sixth of the county total of 620,000). If growth continues along similar lines, even on a gradually diminishing basis, the county population could be expected to top 1,000,000 before 1980. For this reason, it is especially important that sound planning principles be applied when dealing with this growth.

Charlotte County

Charlotte County, the smallest county geographically of the ten-county southern

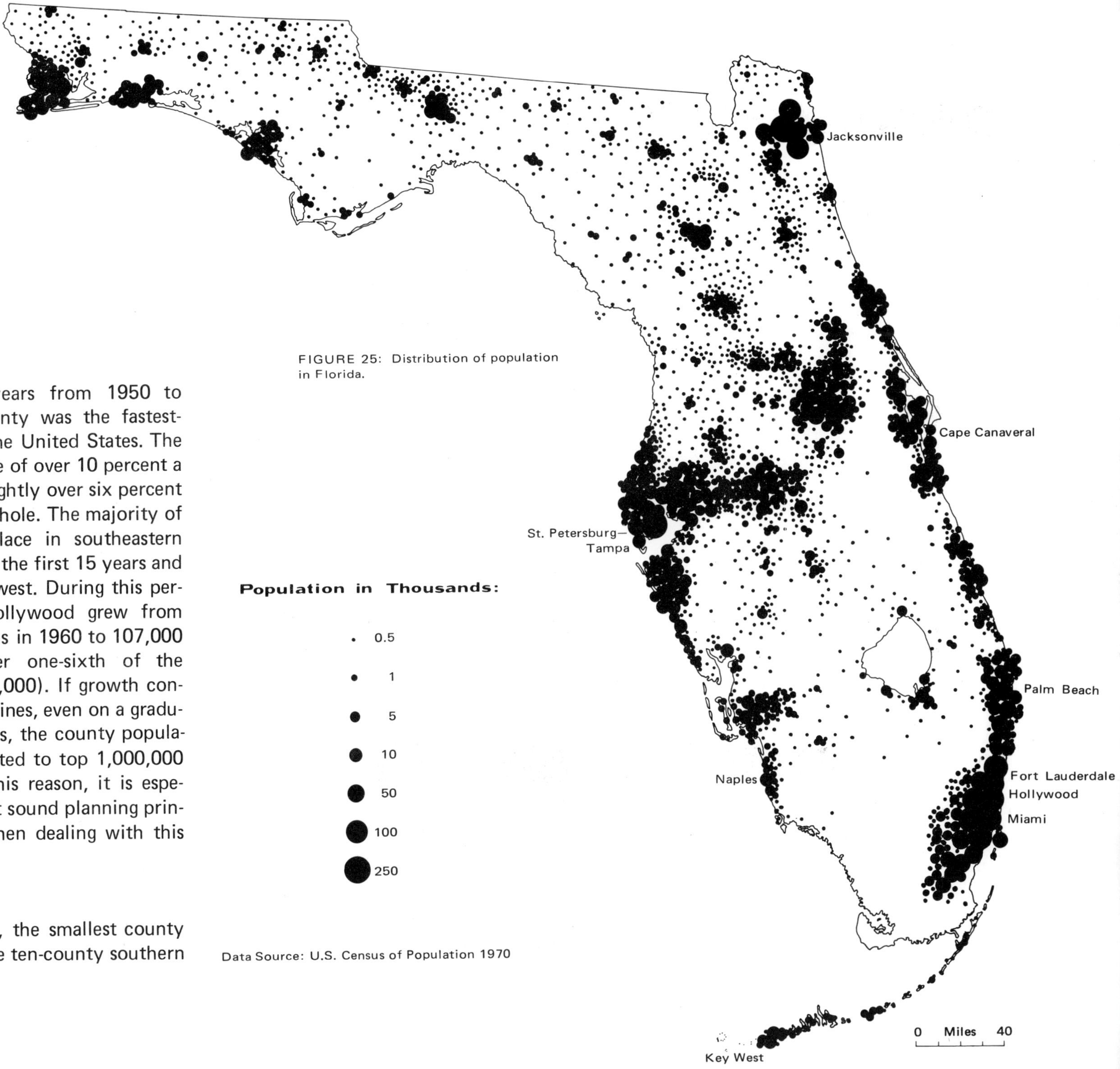

FIGURE 25: Distribution of population in Florida.

Population in Thousands:

- 0.5
- 1
- 5
- 10
- 50
- 100
- 250

Data Source: U.S. Census of Population 1970

Jacksonville

Cape Canaveral

St. Petersburg—
Tampa

Palm Beach

Naples

Fort Lauderdale
Hollywood
Miami

0   Miles   40

Key West

Florida region, lies to the northwest on the Gulf of Mexico. Although this county had a high annual growth rate from 1950 to 1970 of almost 10 percent per year, the actual increase only amounted to 23,300 persons, the difference between 4,300 in 1950 and 27,500 in 1970. The majority of this growth resulted from the development of Port Charlotte, a new community that grew from about 3,000 in 1960 to 10,700 in 1970. During this same ten-year span, Punta Gorda, the county's principal town, grew from 3,000 to 3,900. By mid-1974, Port Charlotte was already topping 23,000 persons, and it has a projected final population of 300,000. Persons over 60 years of age comprise approximately 50 percent of the county's population.

Collier County

Over half of the 2,119 square miles contained within Collier County lie in the Everglades water basin or other environmentally important areas such as the Big Cypress and Corkscrew Swamps and the Fakahatchee Strand. Like Charlotte County to the northwest, Collier County showed an impressive growth rate from 1950 to 1970 (over nine percent annually), but the actual growth in terms of total numbers was small. Collier County's population in 1950 was 6,500 and twenty years later had risen to slightly over 38,000. Predictably, population growth was concentrated in the urbanized areas along or near the Gulf coast. The Greater Naples area contains approximately 75 percent of the population of the entire county.

Collier County is one of the least populated counties of southern Florida, with a population density of 18 persons per square mile. The reserve of developable land, and the fact that the large metropolitan counties to the East (Dade, Broward, and Palm Beach) are filling up, combine to give Collier County a potential for increased population growth. It is expected that the federal and state governments will exercise considerable control in the future development of this area, as indicated by their combined efforts to insure a future water supply by purchasing much of the Big Cypress Swamp.

Dade County

Dade County is the largest metropolitan area in southern Florida as well as in the state. It is also the economic capital or focal point of the region, supplying the area with the economic, social, medical, governmental, educational, and transportation functions typically associated with such a capital. The major growth that the area experienced in the period between 1950 and 1970 was basically a suburban phenomenon. The suburban community of Coral Gables grew from 19,800 in 1950 to 42,500 in 1970, while Hialeah grew from 19,700 in 1950 to 102,000 in 1970. During this same period the already devel-oped city of Miami grew from 250,000 to 335,000, or only 34 percent, as compared to Hialeah's growth of over 500 percent. During the period from 1960 to 1970, Dade County's growth was largely due to the influx of 300,000 Cuban refugees, who have added a distinctive Latin American cultural flavor to the area and have provided a labor force which, among other things, has made Miami the center of the nation's sportswear and playclothes manufacturing industry.

With a population of roughly 1,270,000 in 1970, Dade County is the most densely settled county in the region, with slightly over 600 persons to the square mile. That figure, however, is based on the total area in the whole county, whereas if only the urbanized area is considered, the density figure jumps to ten times that amount, about 6,000 persons per square mile. Other counties have experienced rates of growth comparable to or exceeding those of Dade and Broward Counties, but the population bases were so low that the numerical increases were minimal.

Glades County

Glades devotes the vast majority of its land area to agriculturally oriented activities with only 200 persons out of its total 1970 population of 3,700 engaged in its only industry, the processing and packaging of local agricultural products. Any new development in this county would

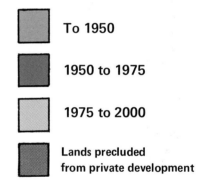

26
projected urban growth

Legend:

- To 1950
- 1950 to 1975
- 1975 to 2000
- Lands precluded from private development

Map labels:

STUART

LAKE OKEECHOBEE

PALM BEACH

FORT LAUDERDALE

SANIBEL ISLAND

NAPLES

MARCO ISLAND

TEN THOUSAND ISLANDS

GULF OF MEXICO

MIAMI

KEY BISCAYNE

ELLIOTT KEY

ATLANTIC OCEAN

FLORIDA BAY

FLORIDA KEYS

KEY WEST

north
mi.
km.

0 5 10 20 30 40

0 5 10 20 30 40 50 60

bring about radical changes in the growth rate and the character of the population.

Hendry County

Hendry, like Glades, is an agricultural county with only about one percent of its total land area given over to urban development. The 1970 census listed Hendry's population as 11,859 persons, three times larger than that of Glades. The 1950-1970 growth rate was less than four percent per year, growing from 6,000 in 1950 to 11,900 in 1970. Most of the county's growth was rural in nature. One of the two principal cities, Clewiston, grew in population from roughly 2,500 to 3,900, while La Belle's population remained essentially the same at 1,800. Industry, employing about 280 persons, consists of the processing of local agricultural products.

Lee County

Lee County, like the Gold Coast counties, was recently designated as a Standard Metropolitan Statistical Area (the Census Bureau's classification for an essentially urban metropolitan area of 50,000 persons or more and exhibiting growth potential). The basis for the county's growth between 1950 and 1970 has consisted mainly of persons over 60 years of age who now comprise over one-fourth of the total population.

Although the population of the county's principal city, Fort Myers, has doubled in the past 20 years, from 12,300 to 27,400, the county's population increased from 23,400 to 105,200 during the period from 1950 to 1970. The majority of this growth was non-urban, taking place in "developer communities" such as Cape Coral and Lehigh Acres. Unless industrial development occurs, the growth rate will probably decline as land values continue to rise and move beyond the financial scope of retirees with fixed incomes.

Martin County

Martin County lies at the northern end of southern Florida's rapidly growing Gold Coast. The county grew eight percent per year between 1950 and 1970, augmenting the base population of 8,000 in 1950 to roughly 28,000 by 1970. The county's major city, Stuart, did not reflect this growth, moving from a base of about 3,000 in 1950 to 5,000 in 1970, although the growth in the area around Stuart and south along the Martin County coast, is, however, now gaining momentum.

Monroe County

Although Monroe County is usually thought of as the area of the Florida Keys, the majority of the county lies on mainland Florida and is almost entirely contained within the boundaries of the Everglades National Park. The population, however, is concentrated in the Keys. Population growth occurred all along the Keys except in Key West, which actually lost population between 1960 and 1970, dropping from 34,000 to 28,000, while county population rose from 30,000 to about 52,500. Retirees constitute a growing percentage of the population, with persons over 60 rising from 7.5 percent in 1950 to 13.3 percent in 1970.

Palm Beach County

If the water surface portion of Lake Okeechobee is included, Palm Beach County is the largest county in the state of Florida, with a total area of roughly 2,500 square miles. Although coastal Palm Beach was the first of the Gold Coast counties to develop because of attractive living conditions, it lacked a port and was quickly exceeded by Miami. The population has been growing most rapidly in the southern area surrounding Boca Raton, which borders Broward County; this growth has been attributed to a "spillover" effect from Broward. After several decades of relatively slow growth, the population of Palm Beach County has risen from about 115,000 in 1950 to 349,000 in 1970.

The Situation Today

All metropolitan areas progress through the same general stages of development in the process of their growth except for "new towns" or planned communities, which develop under rigidly controlled conditions. Except for these planned com-

TABLE 6
POPULATION AND LAND USE TRENDS

1970	Dade	Broward	Palm Beach	Lee	Monroe	Collier	Martin	Charlotte	Hendry	Glades	Totals
TOTAL LAND AREA (square miles)	2,050	1,220	1,980	770	1,030	2,030	560	700	1,190	750	12,280
Area which is government regulated (square miles)	660	470	270	*	670	700	20	*	*	*	2,820
1970 TOTAL POPULATION	1,267,792	620,100	348,753	105,216	52,586	38,040	28,035	27,559	11,859	3,669	2,503,609
Urbanized or built-up area (square miles)	210	140	150	100	50	30	30	*	*	*	725
Persons per square mile	6,040	4,430	2,330	1,050	1,050	1,270	930	1,840	1,480	1,220	3,450
1975 TOTAL POPULATION	1,452,000	923,000	481,000	155,000	59,000	60,000	43,000	30,000	19,000	8,000	3,230,000
Urbanized or built-up area (square miles)	240	200	200	150	50	50	40	20	*	*	980
Persons per square mile	6,050	4,620	2,410	1,030	1,180	1,200	1,080	1,500	1,460	1,140	3,295
1980 TOTAL POPULATION	1,536,000	1,042,000	730,000	175,000	61,000	50,000	64,000	35,000	20,000	16,000	3,729,000
Urbanized or built-up area (square miles)	250	230	210	170	60	50	30	20	*	*	1,050
Persons per square mile	6,140	4,530	3,480	1,030	1,020	1,000	2,130	1,750	1,430	1,230	3,550
1990 TOTAL POPULATION	1,884,000	1,228,000	810,000	197,000	72,000	61,000	74,000	42,000	33,000	28,000	4,429,000
Urbanized or built-up area (square miles)	300	270	250	200	60	60	50	20	20	20	1,250
Persons per square mile	6,280	4,550	3,240	990	1,200	1,020	1,480	2,100	1,650	1,400	3,540
2000 TOTAL POPULATION	2,268,000	1,436,000	940,000	241,000	85,000	72,000	87,000	51,000	48,000	42,000	5,270,000
Urbanized or built-up area (square miles)	370	320	300	240	80	60	60	30	30	30	1,520
Persons per square mile	6,130	4,490	3,130	1,000	1,060	1,200	1,450	1,700	1,600	1,400	3,470

*Less than 20 square miles

NOTE: The purpose of this table is to give a general overview of population and land use trends. Due to rounding, totals may not equal the sum of each column.

SOURCES: U.S. Department of Commerce, 1970; U.S. Department of Agriculture, 1970.

munities, most cities begin as small towns that gradually grow and develop as commerce and industry flourish, attracting more and more people until each community has a distinct downtown or central business district, with the resident population spilling over into suburban areas.

Similar trends were followed all over the nation in the post-World War II rush to the suburbs to escape the congested central cities. Commuting became a way of life, and the massive building boom that ensued was made possible by new highway programs that made the outlying areas accessible and low-cost FHA and VA loans. With this shift in emphasis, compact urban centers gave way to metropolitan sprawl that required the provision of costly municipal services.

New commercial development in the suburbs siphoned off business and reduced the tax base of the central cities, which were already suffering the effects of the exodus of the more affluent residents to the suburbs. In general, only the poor remained in the central cities. Recently, strenuous efforts have been made to restore balance by revitalizing the central core of many large cities; however, the success of these programs remains to be seen.

infra-structure.

The word infrastructure has come into wide use in recent years among planners, and refers generally to the physical installations in any given area that are necessary to support a community. Infrastructure includes roads, rail lines, canals, power lines, ports, and other such facilities, but does not include dwelling units or commercial structures. The infrastructure of South Florida is fascinating in its complexity, and when seen on a map, it is easy to understand how it may well hold the key to the future development patterns of South Florida.

Recent thinking in the field of urban and regional planning is tending toward the "corridor" approach to transportation, the movement of goods, transmission of power, and so on, with feeder-routes connecting the outlying areas of a community to the corridor. This concept is becoming a reality in many areas of the nation as urban populations expand and new living space is created to accommodate them.

South Florida, however, has particular constraints that must be addressed before development can take place in certain areas. As can be seen by referring to several of the maps in this study, large areas of southern Florida have now been brought under government control or supervision. These are areas of "critical concern" to the maintenance of the fragile ecosystem upon which all life in this subtropical area depends.

The three southeastern counties that make up Florida's "Gold Coast," Dade, Broward, and Palm Beach counties, have all grown up to the point that urban development is now pressing closer and closer to their western limits. These "limits" are the dikes or levees of the Flood Control District and represent the farthest point west which urbanization may go in these eastern seaboard counties.

The various components of southern Florida's infrastructure have played a major role in the overall development of the area, and a closer look at these individual elements will help to clarify their impact.

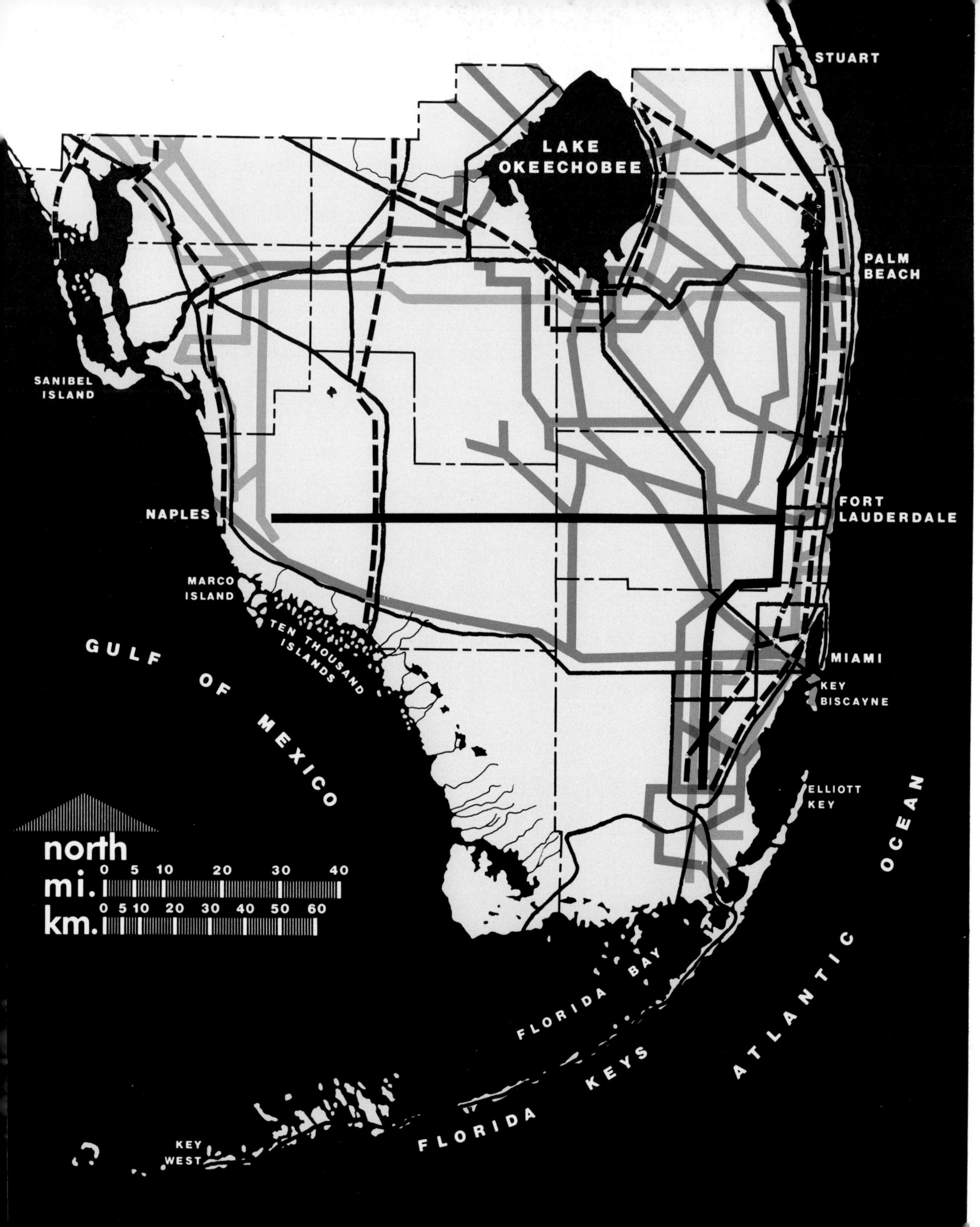

STUART

LAKE
OKEECHOBEE

PALM
BEACH

SANIBEL
ISLAND

NAPLES

FORT
LAUDERDALE

MARCO
ISLAND

TEN THOUSAND ISLANDS

GULF OF MEXICO

MIAMI

KEY
BISCAYNE

ELLIOTT
KEY

north
mi. 0 5 10 20 30 40
km. 0 5 10 20 30 40 50 60

FLORIDA BAY

FLORIDA KEYS

ATLANTIC OCEAN

KEY
WEST

infra-structure 27

	Major highways
	Primary routes
	Railroads
	Canals
	Power lines

SOURCE

"Rand McNally Road Map of Florida," 1974.

Central and Southern Florida Flood Control Project Map, 1971.

Section Maps of Florida, prepared by the Army Map Service (KCSX), Corps of Engineers, U.S. Army, Washington, D.C., 1974.

State of Florida Electric System Map, prepared by Florida Power & Light Co., 1974.

For example, Florida's electrical power is provided by ten generating plants operating in Florida Power and Light Company's four southernmost divisions. All but two (at Sarasota and Fort Myers) are on the southeast coast of Florida, which reflects the concentration of power-consumptive users in this rapidly growing area. Electrical energy consumption in southern Florida follows a different pattern from that of the nation as a whole. Only one-third of the power consumed nationwide is used for residential purposes, but in Florida this figure exceeds 50 percent; and, although national industrial power usage approaches 50 percent, it accounts for less than 10 percent in southern Florida. These figures immediately emphasize the character of the area. Heavy industry typifies the northern urban centers of the nation, but Florida thrives on clean, light industry that is compatible with a tourist-oriented economy and a life-style that is heavily dependent on air conditioning.

Another facet of any area's infrastructure is its transportation complex, and in this region the focus for the movement of vast numbers of people and huge amounts of cargo in and out of the area is Miami's International Airport. This facility is one of the five most active airports in the world from the standpoint of the amount of international air cargo and the number of domestic jet flights per week. This statistic is not difficult to understand when one considers that the Miami area has a transient tourist population in excess of 600,000 persons per month throughout the year, and the majority of them come and go by air.

Southern Florida has two major deepwater seaports for the movement of passengers and cargo. One is located at the Port of Miami, and the other is located approximately 25 miles to the north at Port Everglades in the Fort Lauderdale-Hollywood metropolitan area. The Port of Miami can accommodate oceangoing vessels in its 32-feet-deep ship channel, and Port Everglades can handle deep-draft ocean liners and international cargo vessels with its 37-feet-deep ship channel. Both ports do a brisk business in cruise ships and containerized cargo, and together they offer over a million square feet of warehouse space and several hundred acres of open storage. Port Everglades has traditionally handled a higher volume of bulk cargo (roughly ten times Miami's normal business), whereas Miami regularly handles roughly ten times Port Everglades total in cruise ships; so, these twin neighboring ports complement each other in function and, combined, put South Florida in competition with some of the larger northern ports.

The southern Florida area is served by rail on both the east and west coasts as well as by high-speed, limited-access expressways and interstate highways. Although the rail system does not cross the state south of Lake Okeechobee, two major highways do perform this vital linking function. One is the Everglades Parkway (known locally as Alligator Alley), which runs on a straight east-west path between the Fort Lauderdale-Hollywood area on the east and Naples on the west. The other highway performing the same function is U.S. 41 (or the Tamiami Trail), which originates in Miami and also connects with Naples on the west coast.

A brief study of the map shows large areas in the central portion of southern Florida that appear to be readily developable. Taking into consideration lands under government ownership or control and areas of critical concern from the standpoint of South Florida's freshwater supply, an area covering over five thousand square miles of land in the center of South Florida must, for all practical purposes, be considered currently undevelopable.

Furthermore, with the cost of utilities and municipal services constantly increasing in an inflationary economy, the prospective builder-developer must look to existing roads, power lines, railroads, and communications networks before selecting a site. If he does not, local government may place constraints on his development plans in order to avoid being forced into providing utilities and public services for outlying developments over distances that become prohibitively expensive for the municipality involved.

recreation

One of the basic reasons for the constant influx of people to southern Florida is recreation and the total life-style that revolves around it. This total life-style incorporates climate, outdoor activities, and a subtropical setting that provides opportunities for everything from collecting rare butterflies in the shady recesses of the Big Cypress Swamp to photographing tiny, rare, tropical fish in the only underwater parks in the continental United States.

These amenities of life and the ever-rising standard of living have had the very practical side effect of making leisure-time and its associated recreation activities among the fastest growing industries in the nation. Thus, a major segment of American industry is now devoted either directly or indirectly to providing the goods and services that an expanding segment of the population desires for maximum enjoyment of leisure-time activities. Automotive companies produce recreation vehicles that range from the most modest camper to a "penthouse" on wheels; other assorted recreational vehicles are also produced, including dune buggies, swamp buggies, and airboats. Southern Florida is a mecca for boat owners, and every marina in the study area, indeed over the entire state, usually has waiting lists for dock space. Recreational boating has come of age since World War II, and now the ownership or rental of some type of watercraft is within the financial reach of almost anyone.

In 1965, the Florida Outdoor Recreation Development Council conducted a survey of Florida residents in an attempt to determine participation in selected outdoor recreation acitivities. Survey data were collected in both winters and summers of 1964 and 1965. Beach activities proved to be the most attractive form of recreation to both tourists and residents, followed by picnicking and saltwater swimming (see Table 7). Since one person may participate in more than one activity, the ranking reflects participation per person in each activity.

The importance of recreational activities in the southern Florida life-style is reflected in the design of new residential communities where such facilities as swimming pools, shuffleboard courts, paddle ball courts, saunas, and game rooms form an essential part of the total concept of life there. In fact, entire residential complexes have been developed around a single sport, of which golf, tennis, boating, and horseback riding are examples.

Thus, the recreational opportunities in South Florida have strongly motivated many people to reside in this area permanently. In the past the trend has been for retirees and people who need a warm climate for health reasons to relocate here. Recently many international corporations have located regional home offices in this area to better serve their Caribbean and Latin American operations; employees find the recreational possibilities an enticement to live here. Recreation in all of its diversity therefore plays a preeminent role in the attraction of tourists and residents alike.

Finally, it is interesting to note that one of the greatest movements of people in history occurs regularly every year when over twenty million people come to vacation in Florida. Because southern Florida hosts between five and ten million of these visitors, it is not difficult to understand the tremendous impact that tourism has on the economy of the area.

TABLE 7

OUTDOOR RECREATION PARTICIPATION BY RESIDENTS AND TOURISTS
SOUTHERN FLORIDA STUDY AREA

Activity	Resident population Percent	Resident population Rank	Tourist population Percent	Tourist population Rank	User Occasions Residents	User Occasions Tourists	User Occasions Total
Beach activities	61.4	1	44.0	1	21,063,000	13,100,500	34,163,500
Picnicking	55.5	2	17.5	5	9,271,500	4,200,000	13,471,500
Swimming in salt water	45.5	3	36.0	2	17,496,500	11,028,500	28,525,000
Visiting historical sites	29.8	4	23.4	4	1,172,500	5,775,000	6,947,500
Fishing in salt water	29.4	5	26.3	3	8,890,000	8,396,500	17,286,500
Fishing in freshwater	27.8	6	8.7	8	9,306,500	2,516,500	11,823,000
Swimming in freshwater	27.5	7	10.4	7	5,841,500	2,887,500	8,729,000
Boating	24.1	8	13.8	6	9,898,000	3,948,000	13,846,000
Camping	18.1	9	5.5	10	1,473,500	1,400,000	2,873,500
Water skiing	10.6	10	4.7	11	3,174,500	973,000	4,147,500
Nature study	10.0	11	5.7	9	3,132,500	1,802,500	4,935,000
Hunting	7.5	12	*	13	1,337,000	*	1,337,000
Hiking	4.5	13	3.6	12	588,000	787,500	1,375,500
Total					92,645,000	56,815,500	149,460,500

*Insignificant

SOURCES: Report for Kissimmee-Everglades Area, U.S. Department of Agriculture, in cooperation with the Division of Interior Resources, Florida Department of Natural Resources, 1973.

Florida Outdoor Recreation Development Council, "Outdoor Recreation in Florida, 1965, "Tallahassee, Florida, 1968.

Department of Commerce, State of Florida, "Florida Tourist Survey, 1964 and 1965," Tallahassee, Florida.

RECREATION

1. Long Pine Key
2. Flamingo
3. Royal Palm Hammock
4. Rookery Bay Sanctuary
5. Fisheating Creek N.W.R.
6. Island Bay N.W.R.
7. Caloosahatchee River N.W.R.
8. Matlacha Pass N.W.R.
9. Jay N. "Ding" Darling N.W.R.
10. Pine Island N.W.R.
11. Key West N.W.R.
12. Great White Heron N.W.R.
13. Key Deer N.W.R.
14. Loxahatchee N.W.R.
15. Hobe Sound N.W.R.
16. Collier Seminole S.P.
17. John Pennekamp Coral Reef S.P.
18. Jonathan Dickinson S.P.
19. Saint Lucie Inlet S.P.
20. Copeland Fire Tower
21. Koreshan S.R.A.
22. Pahokee S.R.A.
23. Bahia Honda S.R.A.
24. Long Key S.R.A.
25. Cape Florida S.R.A.
26. Hugh Taylor Birch S.R.A.
27. Port Charlotte Beach S.R.A.
28. Caloosahatchee River S.R.A.
29. Wiggins Pass S.R.A.
30. Cowpens Rookery, State Preserve
31. Okee-Tantie Recreation Area
32. Belle-Glade Recreation Area
33. Ortona Lock Recreation Area
34. Saint Lucie Lock and Dam Recreation Area
35. Cecil C. M. Webb. W.M.A.
36. Fisheating Creek W.M.A.
37. J. W. Corbett W.M.A.
38. Aero Jet W.M.A.
39. Everglades W.M.A.
40. Brown's Farm W.M.A.
41. Cape Haze Aquatic Preserve
42. Pine Island Sound Aquatic Preserve
43. Cape Romano Aquatic Preserve
44. Coupon Bight Aquatic Preserve
45. Biscayne National Monument
46. Estero Bay Aquatic Preserve
47. Intracoastal Waterways Aquatic Preserve
48. Fort Jefferson National Monument
49. Indian Key Historical Site
50. Herrera Sunken Ship, Historical Site

W.M.A. - Wildlife Management Area
N.W.R. - National Wildlife Refuge
S.R.A. - State Recreation Area
S.P. - State Park
I.R. - Indian Reservation

LAKE
OKEECHOBEE

STUART

PALM
BEACH

FORT
LAUDERDALE

MIAMI

KEY
BISCAYNE

ELLIOTT
KEY

SANIBEL
ISLAND

NAPLES

MARCO
ISLAND

TEN
THOUSAND
ISLANDS

GULF OF MEXICO

FLORIDA BAY

FLORIDA KEYS

ATLANTIC OCEAN

KEY
WEST

north
mi.
km.

0 5 10   20   30   40

0 5 10   20   30   40   50   60

A
B
C

○ Private Camps and Trailer Parks

National Wildlife Refuge
Wildlife Management Areas

Big Cypress National Preserve

Everglades National Park

A. Brighton Seminole Indian Reservation
B. Big Cypress Seminole Indian Reservation
C. Florida State Seminole Indian Reservation

FCD Conservation Area

Audubon Society Corkscrew
Swamp Sanctuary

▲ Recreational facilities within Everglades Conservation Areas

★ Primitive Campsites within Everglades National Park

SOURCE

"Rand McNally Guide to Florida," 1973.

"Land Use & Highway Functional Classification Systems," 1970, Wilbur Smith & Associates.

Recreational Map of the Everglades, Central
and Southern Florida Flood Control District,
1974.

Recreation Guide to Lake Okeechobee, Central and Southern Florida Flood Control
District, 1973.

florida's past

Florida's history stretches back some ten thousand years in time. During this period, the Calusa Indians who inhabited the southern part of the peninsula repelled Spanish settlement in the sixteenth century following Ponce de León's voyage of discovery in 1513. Spain finally established a number of missions and forts and introduced European "culture" to the area, but true development of the state was not to come for several centuries.

In the intervening time span, Florida was given to England by Spain in 1763 in exchange for Havana, and the territory remained loyal to England during the American Revolution. British loyalists streamed to Florida from the rebellious colonies, and John Hancock was even burned in effigy in Saint Augustine. Twenty years later, in 1783, Florida was given back to Spain by England in exchange for the Bahamas and Gibraltar, and it remained in Spanish hands for another twenty years until 1803 when the United States claimed the Florida Panhandle following the Louisiana Purchase and a series of border disputes with Spain and France.

During the Napoleonic Wars, which engulfed Europe for years and finally involved the United States in what we now refer to as the War of 1812, Florida's status was uncertain; Florida finally was acquired in 1821 by the United States following the First Seminole War.

Andrew Jackson was made the first governor of the Florida territories following his military victories over the Indians, and in 1845 following the end of the Second Seminole War, Florida became a state with some 66,000 inhabitants. Although Florida seceded from the Union in 1861, the state was relatively untouched by the Civil War, and for the next fifty years, a slow but steady trend toward growth began to occur as the railroad moved farther and farther south on its way toward opening up the entire state to development. Men like Henry Plant and Henry Flagler built luxury hotels to accommodate the influx of tourists that their railroads carried to South Florida's tropical seacoast. Hamilton Disston supervised the drainage of lands around Lake Okeechobee for agriculture, and by 1896 his engineers had drained 4 million acres, some of which are still in production today. Even so, there were considerably less than one million inhabitants in the entire state at the end of World War I, and it was only after World War II and the onset of widespread population mobility that the state began to grow dramatically.

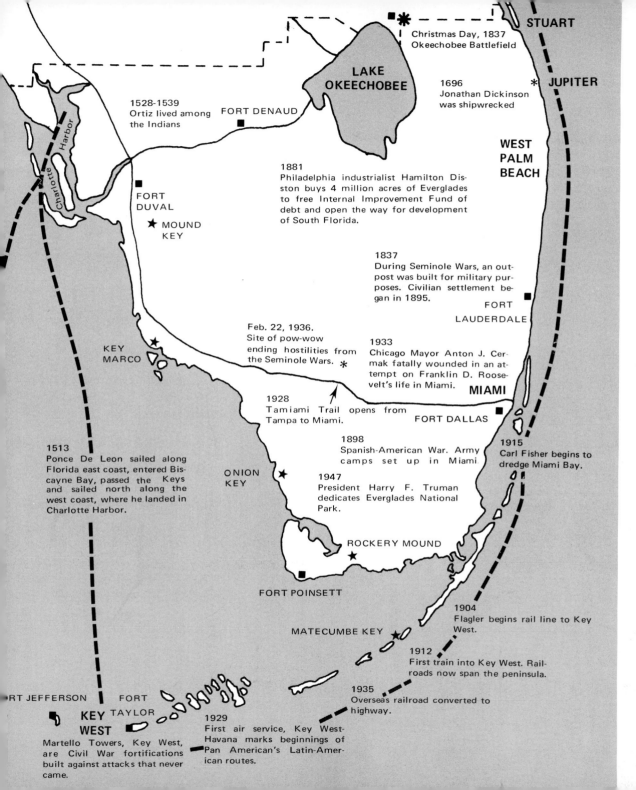

Christmas Day, 1837
Okeechobee Battlefield

LAKE OKEECHOBEE

1696
Jonathan Dickinson
was shipwrecked

1528-1539
Ortiz lived among
the Indians

FORT DENAUD

WEST PALM BEACH

STUART

JUPITER

1881
Philadelphia industrialist Hamilton Disston buys 4 million acres of Everglades to free Internal Improvement Fund of debt and open the way for development of South Florida.

FORT DUVAL

★ MOUND KEY

1837
During Seminole Wars, an outpost was built for military purposes. Civilian settlement began in 1895.

FORT LAUDERDALE

Feb. 22, 1936.
Site of pow-wow ending hostilities from the Seminole Wars. ✳

1933
Chicago Mayor Anton J. Cermak fatally wounded in an attempt on Franklin D. Roosevelt's life in Miami.

MIAMI

KEY MARCO

1928
Tamiami Trail opens from Tampa to Miami.

FORT DALLAS

1513
Ponce De Leon sailed along Florida east coast, entered Biscayne Bay, passed the Keys and sailed north along the west coast, where he landed in Charlotte Harbor.

ONION KEY

1898
Spanish-American War. Army camps set up in Miami.

1915
Carl Fisher begins to dredge Miami Bay.

1947
President Harry F. Truman dedicates Everglades National Park.

ROCKERY MOUND

FORT POINSETT

MATECUMBE KEY

1904
Flagler begins rail line to Key West.

1912
First train into Key West. Railroads now span the peninsula.

1935
Overseas railroad converted to highway.

RT JEFFERSON   FORT TAYLOR

KEY WEST

1929
First air service, Key West-Havana marks beginnings of Pan American's Latin-American routes.

Martello Towers, Key West, are Civil War fortifications built against attacks that never came.

29 florida's past

- ■ Fort
- ✳ Major Battlefield
- ★ Major Archaeological Site
- ✳ Point of Interest
- – – Ponce de Leon's Journey

SOURCE

"Florida Almanac"

Charlton W. Tebeau, "A History of Florida" and personal interview

Marjory Stoneman Douglas, "Everglades, River of Grass" and personal interview

D. Graham Copeland, Map of Collier County, 1947

Military Map of Florida, 1856

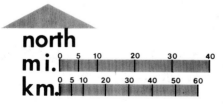

north
mi.
km.

0 5 10 20 30 40
0 5 10 20 30 40 50 60

63

Southern Florida has retained its image as a pleasant subtropical vacation and retirement center, and funds from these and related supporting service activities make up a large segment of South Florida's economic base.

South Florida's image has been so deeply established in the minds of the majority of the nation's population that continued growth is almost inevitable. Although Florida is historically young when compared to the development of other areas of the United States, few places have come into national prominence as quickly or as dramatically as South Florida. The area's romantic past, reflected in the remains of significant Indian settlements, the era of the railroads, and the establishment of Everglades National Park and the Big Cypress Preservation Area, adds another dimension to the richness and diversity of the future of the area.

The Federal Census of 1950 reported just over 2,700,000 people in Florida, with the majority of them settled in four or five of the state's larger cities. In the same year, 4,700,000 tourists visited the area, and these pleasure-oriented migrations had grown to over 23,000,000 visitors per year by 1970.

part two

2

the natural environment

the natural environ- ment

Large areas of southern Florida are well suited for urban development. Other areas can be utilized most efficiently as support facilities. The purpose of Part Two is to describe the major ecological regions, their characteristics, structure, dynamics, and particularly, their function in supporting human life and the economic activity upon which human existence depends.

Natural systems supply, transport, treat, and store water; modify the climate; oxygenate and purify the air; produce food; treat or assimilate waste; build land; maintain beaches; and provide protection from hurricanes. These systems also provide a unique and beautiful environment that has attracted visitors and residents to southern Florida at unprecedented rates. No human system yet devised to provide these services works as efficiently, dependably, or cheaply as a natural system. Like any other complex system, environmental systems can operate efficiently only as long as they remain in a functional state. If essential compo-

nents are destroyed, or if the system as a whole is overstressed, the process will break down and the system will fail. Therefore, it makes economic sense to utilize each system for the function to which it is best suited and to maintain peak efficiency by intelligent management.

Environmental Services

One way of evaluating natural systems is according to the type of support services they provide for human activity. This does not imply their other functions are unimportant—many operate on a global scale—but this discussion is concerned with natural systems at a regional or subregional level, closer to human concern.

Primary Life Support. These are the services that only natural systems can reasonably provide for the maintenance of human life. Examples are: production of food (carbohydrates) directly from carbon dioxide, water, and energy from the sun through photosynthesis, or the recycling of water in the hydrologic cycle.

Secondary Life Support. These services are provided by natural systems at no cost, but can be partly or wholly duplicated by human systems at a cost. Sewage treatment is one example.

Economic Support. This term applies to a natural resource, advantage, or amenity that sustains or enhances human economic activity. For example, the climate and environmental uniqueness of southern

Florida attract tourists and new residents, thereby promoting economic growth. Commercial and sport fishing depend in part on the efficiency of marsh and estuarine systems that provide food and shelter for juvenile and adult stages of many species.

Quality of Life. There are two components to quality of life. One is the standard of living, which consists of the services and amenities provided by technology and the built environment; the other is environmental quality, or the services and amenities of the natural environment. An increase in one component often results in a decrease in the other. When the environmental quality component is plentiful and the standard of living is low, greater value is placed on increased consumption to raise the standard of living. This increases the quality of life. As the technological outputs of resource depletion and pollution make environmental quality scarce, however, its value increases. Consequently, a further increase in consumption becomes less important. Most humans have no use for three refrigerators or ten television sets. An increase in environmental quality then becomes more valuable than an increase in standard of living. The fast-growing demand for outdoor recreation opportunity indicates that this point has been reached in American society. Other societies that have not yet achieved a high standard of living will continue to choose increased consumption.

Carrying Capacity

Bohnsack (1974) defines carrying capacity of an environment as "the population size that can be supported in a region at a given standard of living for a defined period of time. The carrying capacity is determined by land area, availability of water and mineral resources, fertility of soils, and the ability of biological systems to assimilate refuse without breakdowns that reduce essential services." He points out that carrying capacity is not fixed, but changes as conditions change. It can be deliberately increased by increasing the inputs of energy (money) or by lowering the standard of living. Carrying capacity decreases as the standard of living rises, because greater demands are placed on the environment as a source (supply or resources, land and water) and as a sink (waste disposal to the air, water, and land surface). These demands are often mutually exclusive. A source of potable water, for example, cannot also be used as a sink for sewage waste. One use or the other will suffer.

Conflicting uses and increasing demands place a stress on the system. In terms of stress, carrying capacity might be defined as the capacity of a system to support a given level of stress without breaking down. Breakdown automatically reduces both the standard of living and environmental quality because carrying capacity of the entire system or of some

environmental component of the system has been exceeded. The component that breaks down first is called the limiting factor (Bohnsack, 1974).

There are several ways to deal with stress. The most obvious and easy solutions are not necessarily the best. In a complex system, solution of a problem in one sector may cause another problem elsewhere—often exactly when it is least expected and most costly. This is especially true if the system has many components, because actions become amplified as they affect each component and results are often the exact opposite of the expected outcome.

Reducing Stress. The most direct and obvious solution is to reduce stress, either by reducing the standard of living (per capita consumption of land, water, etc.) or by slowing population growth. Since the first alternative seems unlikely to win public acceptance, it is "no-growth" that has been widely used as a tool to prevent system breakdown. Temporary moratoriums can buy time for society to develop alternative strategies for dealing with stress. Permanent moratoriums, however, have important impacts on other system components. The natural environment is affected positively, at the expense of the economic component. If slowing of housing construction is selected as the strategy for slowing growth, the price of housing is driven up, while job opportunities decrease and unemployment rises. The result

is a reduction in the standard of living at low and middle income levels, rather than at all levels. Thompson (1973) calls this the "quick and dirty" approach to growth control, and the courts seem to agree that it is exclusionary. It is possible to combine growth control with other strategies.

Increasing System Capacity. Capacity for supporting the growth of population and a rising standard of living can be increased by raising the capacity of the total system to withstand stress. Such a move requires a greater input of energy (money) into the system. Examples of this approach are the application of chemicals to farmland to increase production or the building of sewage treatment plants instead of septic tanks to increase capacity for waste treatment. Often a new technology is needed to increase system capacity without decreasing quality of life. Although this strategy will be effective for a time, eventually a limit will be reached.

The first component of the system to break down is the first limiting factor. If technology can be found to reduce stress on that component, carrying capacity is increased until the second limiting factor is reached. At some point the appropriate technology will not be available to overcome a limiting factor, and either population will stabilize or standard of living will fall. Quality of life, however, is usually degraded in any case as technologic solutions are applied. For example, money can be spent to construct and maintain seawalls, which increases buildable land area to support a larger population, but only by trading off the recreation opportunities and aesthetic appeal of the natural beach.

If a technologic limit is not reached first, economic limits will act to make further increase of system capacity infeasible. Development technology is an example. Government regulations are aimed at reducing stress on environmental systems or infrastructure. Compliance with the requirements imposes an economic cost that may either price the development out of the reach of the largest markets or may make it altogether infeasible. The costs are paid by decreasing the standard of living of the developer, the consumer, and the labor force, but environmental quality is maintained. If technology were available at reasonable cost, a trade off between economic and technologic solutions could be made without degrading the quality of life.

Balancing the System. A third approach to dealing with stress would be a planning and management strategy designed to maximize the efficiency of environmental services. The basic requirement is to apply the stresses in rates, forms, and locations such that the system can absorb and process them without an overload occurring. At the same time, life-support functions and natural amenities can continue to enhance the quality of life.

The first step is to understand how the natural systems function. Ecologists have been successful in studying and quantifying many of the complex processes. New techniques are available for putting a value on environmental services. For example, Gosselink, Odum, and Pope (1973) calculated the value of a tidal marshestuary on the basis of a number of functions performed.

"One very important contribution estuaries make to the growth and economic wealth of highly urbanized regions is the waste treatment that active ecosystems can accomplish without appreciable reduction in water quality. . . . In general, the sewage discharge in these [mid-Atlantic] estuaries does not stop at the 'secondary' stage, but continues through the 'tertiary' stage of nutrient removal and assimilation. Since artificial tertiary treatment of sewage is very much more expensive than secondary, then an acre of marsh-estuary is doing about $14,000 worth of work per year at a daily loading of 19.4 lb. BOD, assuming the cost of artificial tertiary treatment is at least $2/lb. BOD. In other words, this is what it would cost man to deal directly with his wastes if the acre was not available to do this work. Resorting . . . to the income capitalization calculation an acre of estuary that is able to handle [this] waste loading . . . is worth a whopping $280,000. It is no wonder that large cities and industrial complexes tend to be located where large bodies of water are available for 'free' treatment plants!

"Of course, it is apparent that mid-Atlantic estuaries are now overloaded to the extent that . . . water quality aspects are reduced to an undesirable level, especially in terms of fisheries and recreation. The value of $280,000/acre thus represents a large 'overload' of work that has serious pollution side-effects, and if continued or increased could result in system breakdown. If the BOD load can be reduced, these estuaries would function better as tertiary treatment plants and be more valuable overall." The authors go on to evaluate several additional economic and life-support functions of the marshestuary (Gosselink, Odum, and Pope, 1973).

Other natural systems perform equally valuable services. If these natural systems are destroyed, then economic inputs (money, energy, technology) are necessary to avoid degrading the standard of living and quality of life. To maintain the economic and amenity services of natural systems at peak efficiency, management at the regional level is necessary. Outputs of waste should be placed in quantities and locations where the waste can be processed by the natural system without overload. In some cases, the disposal may be direct (septic tanks, sanitary land fill) where population density is not great. To increase the carrying capacity to accommodate higher population density, some degree of technologic input, such as sewage treatment, is necessary.

Resources should be withdrawn from the environment at rates that ensure re-

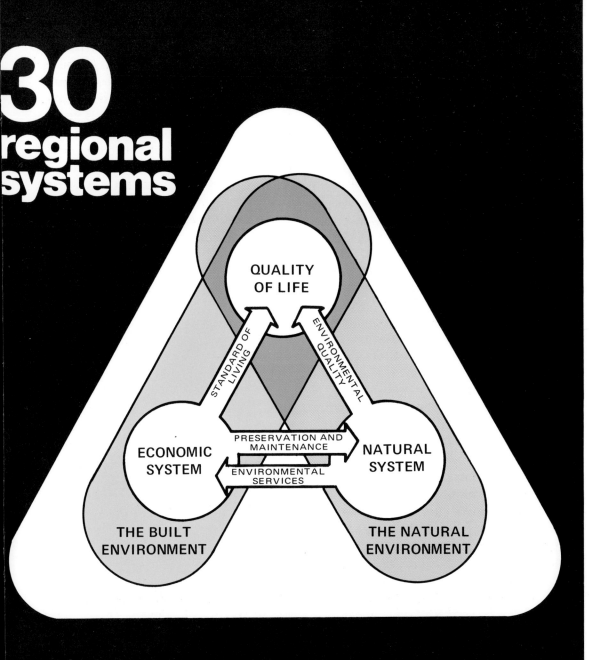

QUALITY OF LIFE

STANDARD OF LIVING

ENVIRONMENTAL QUALITY

ECONOMIC SYSTEM

NATURAL SYSTEM

PRESERVATION AND MAINTENANCE

ENVIRONMENTAL SERVICES

THE BUILT ENVIRONMENT

THE NATURAL ENVIRONMENT

newal and that will not exceed system capacity. Freshwater is the most obvious example. If consumption exceeds the storage capacity of the aquifer, salt intrusion occurs, in which case either water use must be curtailed—a decrease in the standard of living—or desalinization of seawater becomes necessary, requiring greater expenditure of energy (money). Furthermore, overuse of water results in drought and fire in the wetlands, which in turn produce air pollution and decreased recreational opportunities.

Distribution of human activity, i.e., the built environment, should occur in locations that will not destroy valuable natural systems. Densities should be adjusted to avoid putting undue stress on environmental services. Energy-intensive technology may be available to offset the impacts and raise system capacity, but such inputs will also increase the cost of the built environment.

Within the built environment, natural amenities should be preserved because they increase the quality of life and because they have economic value. Trees, for example, are self-maintaining natural units that absorb air pollutants, produce oxygen, cool the air through evaporation and by providing shade, reduce noise, stabilize the soil, and provide aesthetic value both by their own beauty and by providing habitat for other life forms such as orchids or birds. Trees also increase the market value of a building site. Planning for wooded sites should include selective clearing of vegetation and clustering of built units to retain maximum value added to the development by natural vegetation. Replacement cost should be considered before trees are removed, since many native trees are expensive, if available in nurseries. Trees that are in the way of construction can be moved to other sites, or kept in an on-site nursery for later replanting.

During the planning stage, consideration should be given to techniques that are imitative of natural systems. These are usually less costly and more efficient. Storm water, for example, can be diverted to green swales and holding ponds before it is released into a water body. Imitative of a natural wetland, this allows time for vegetation to take up pollutants, and some or all of the water to percolate into the ground to recharge the aquifer. It is also less costly than curbs, gutters, and storm drains.

Development of Marginal Lands

According to a survey of Dade County, omitting the City of Miami, only 50 percent of the available building sites (including many larger parcels) have actually been built upon. The projected Dade population through the year 2000 could be accommodated in areas that are already zoned for use in the presently urbanized area and at the fringe. Although data are lacking for other counties, the 50 percent figure should be on the conservative side,

since Dade has been developed longer and more intensively than any other southern Florida county (Holzheimer, 1975).

Building can earn as much profit on construction in these areas as in outlying development. Taking into account savings of transportation and time costs for both materials and labor, as well as management, potential construction profits may well be greater. Infrastructure costs should be much less. Soil conditions and flood criteria are often more favorable than in marginal outlying areas. For the consumer, a close-in location is worth a higher price. Generally, any savings on housing in outlying areas is offset almost exactly by increased transportation costs. The county tax base will increase irrespective of the ultimate location of a given dwelling unit built within its boundaries, but its costs to service that unit will be lower if it is close to existing services.

If locations within the existing urbanized area can accommodate expected growth, are more marketable, less costly to build on, and less costly for government to service, it makes little economic sense to build in outlying areas. The strategy for land conversion in marginal environmentally sensitive areas is to show a "need" for the proposed housing or other facilities. No attempt is made to evaluate the possibility of accommodating that need in another, more appropriate location. The speculative profits in development of marginal lands are made at the direct expense

of the construction industry, the labor force in that industry, the ultimate consumer, and the increased per capita expenses of local government. This is in addition to indirect costs such as loss of environmental quality, destruction of agricultural lands, deterioration (and subsequent higher cost of treatment) of water resources, and destruction of nearby open space amenities important to the quality of urban life.

Private Land and Public Benefit

When the services of environmental systems benefit their owner, there is an incentive to preserve them for their economic return. Many natural systems, however, may return little or no direct benefit to their private owner, no matter how high their value to society as a whole. The market system fails to work in this case.

"The time has come to seek a means of letting owners of natural resources with high value to society receive a return. The best long term solution is a land use plan which delimits the amount and location of natural areas that will be necessary to support a future optimum level of urban-industrial development. Then such natural areas can be acquired, or zoned, before the spiral of land speculation raises the market price" (Gosselink, Odum, and Pope, 1973).

the coastal strip

The earliest pioneers in South Florida wisely chose the ridge of the coastal strip for their settlements. Here they found refuge from flooding, relief from mosquitoes, and a view of the sea. The trend continues, as 75 percent of the residents of Florida live and work in coastal counties. Because of its higher elevation, superior geologic and soil conditions, and proximity to the sea, the coastal strip remains most desirable and suitable for intensive urban development. Although its intrinsic suitability will remain, certain constraints must be accommodated if the amenities and marketability of development are to be maintained or enhanced, and if the environmental hazards are to be dealt with at least possible cost.

Geology

When the Pamlico Sea stood some thirty feet higher than the present Atlantic, southern Florida was submerged, except for an "island" in the Immokalee area. Wave action formed an offshore bar that would become the coastal ridge once the water receded with the coming of another ice age. North of Boca Raton on the east coast, and Naples on the west, the ridge is comprised primarily of loose sand and shells, in some places cemented together to form "coquina" rock. This cementing is more common in northern Florida, whereas in southern Florida the material is usually unconsolidated. Geologists call this the Anastasia formation. Because of its permeability and elevation, rainfall tends to quickly penetrate or run off, leaving the surface dry. Wells yield water in varying quantities, depending on subsurface conditions. In some places yields are good. Load-bearing capacity is good, since there is almost no marl or organic soil present.

South of Boca Raton the Anastasia formation grades into oolitic limestone, also of marine origin. The ridge becomes wider and lower south of Miami and is cut through by transverse glades. The surface of the formation is rough and pitted, and many solutions holes, both at and beneath the surface, are created by the ongoing action of the acidic surface water as it percolates through the porous rock. Some of these cavities become filled with sand, or with accumulations of organic soil, while others remain open. This structure gives good water-bearing properties to the formation, which acts somewhat as a sponge,

31
the coastal strip

1. **Tree island Everglades:** slightly higher islands of trees in a "sea" of saw grass marsh. Bayheads, willow heads, tropical hammock forests populate the islands.

2. **Wet prairies:** saw grass, maidencane, rushes. Wet most of the year.

3. **Dry prairies:** saw palmetto, grasses, sedges. May be wet part of the year or may remain dry.

4. **Low hammock:** water oak, sabal palm, bay, shrubs. Often occurs where soil is deeper and composed of sand.

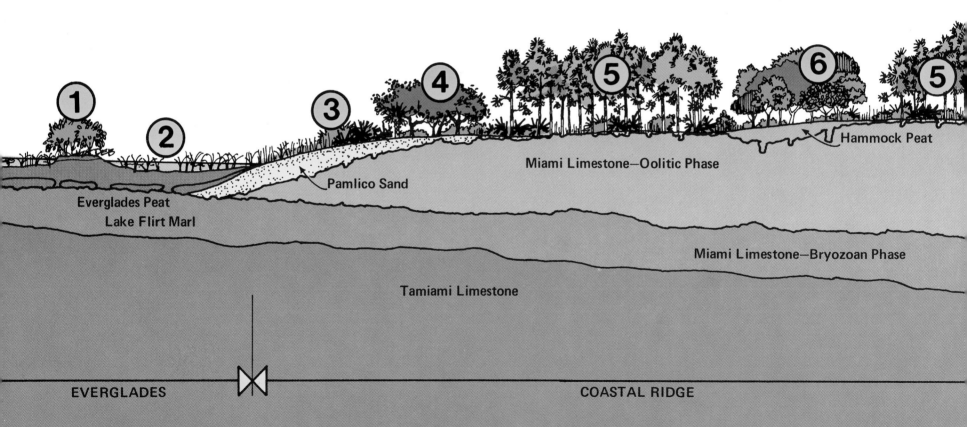

Hammock Peat

Miami Limestone—Oolitic Phase

Pamlico Sand

Everglades Peat
Lake Flirt Marl

Miami Limestone—Bryozoan Phase

Tamiami Limestone

EVERGLADES   COASTAL RIDGE

SOURCES: Hoffmeister, 1974; Long and Lakela, 1971; Parker et al., 1955.

5 Pine-palmetto forest: slash pine, saw palmetto, sabal palm, grasses, and shrubs. Intermittent fires eliminate invading hardwoods, but the fire-resistant pines and palmettos survive.

6 High hammock: tropical hardwoods - lysiloma, mahogany, strangler fig, pigeon plum, gumbo limbo, poisonwood.

7 Coastal marsh: spike rush, cattail, saw grass, sabal palm, small red mangrove, algal mat.

8 Salt marsh: salt grass, cord grass, saltwort, glasswort, scattered small red mangrove.

9 Coastal mangrove forest: red mangrove, black mangrove, white mangrove, buttonwood.

10 Estuary: shallow water, less saline than seawater because of freshwater from upland runoff and percolation from the aquifer. Shoal grass; turtle grass; large, bottom-anchored, and drifting algae (seaweed).

11 Coastal strand forest: sea grape, coconut palm, beach cedar.

12 Beach and dune: sea oats, beach grasses, goatsfoot, morning glory, scaevola, beach lavender.

13 Offshore: reef corals, patch corals, soft corals, branching corals, sea grasses; drifting and attached algae in shallower waters.

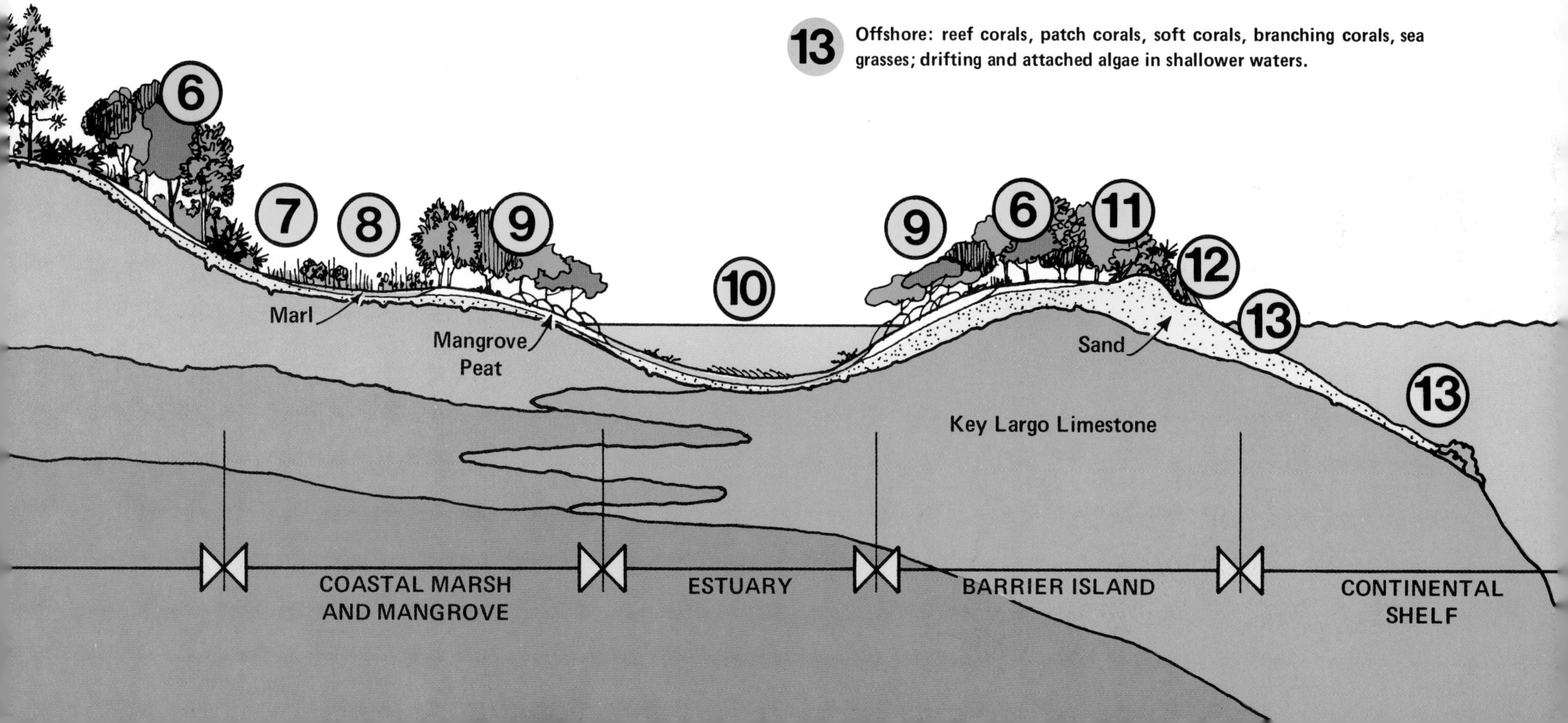

Marl

Mangrove Peat

Sand

Key Largo Limestone

COASTAL MARSH AND MANGROVE

ESTUARY

BARRIER ISLAND

CONTINENTAL SHELF

Storm erosion

Storm damage

Groins on an eroding beach

but it can present localized problems for heavy construction. Borings should be taken to be sure that there are no cavities present.

Topography

The coastal ridge parallels both the eastern and western shorelines of southern Florida. On the east coast, where it is better developed, the ridge originates near St. Augustine and extends southward through the study area to Florida City, with outcrops in Everglades Park. On the west coast it is well defined as far south as Naples, merging with the sandy flatlands on the north. From a maximum elevation of about 20 feet, the ridge slopes almost imperceptibly inland, gradually giving way to the lower and wetter interior regions, the sandy flatlands in the north, and the Everglades and Big Cypress Swamp in the south. Toward the coasts, the slope is more rapid, and the ridge is fringed with mangroves, estuaries, coastal beaches, and

barrier islands before it dips into the sea. It is not a striking feature, but merely a higher place in an essentially horizontal landscape. Except for a few rivers, the surface water flow in the interior Everglades and Big Cypress basins is impounded and diverted southward by the ridge. Historically, in times of flood, water crossed the ridge through ancient tidal channels now known as transverse glades. Many of these have been deepened into permanent drainage canals.

Coastal Strand and Dunes. Sand beaches and dunes occupy a very small part of the southern Florida land area, but they are so attractive and desirable that they have become the symbol of the regional way of life. They form a narrow band along the shore, mostly on the seaward side of barrier islands, and behind them are found the lagoons and wetlands of quieter waters. Beaches are built and rebuilt by the forces of ocean currents, storms, and tides. They are essentially temporary,

since the forces that built them can also alter them beyond recognition—sometimes during a single great storm. Even the most costly stabilization projects have proven ineffective against the forces of beach evolution.

SAND TRANSPORT. Individual grains of sand are constantly being transported by currents from their source areas and moved along the beach. The pattern of movement in the Gulf of Mexico is not well defined, but along the Atlantic shore the longshore current is the dominant sand transport mechanism. Although there is some seasonal variation in the direction of this littoral drift, the net effect is southward, with as much as 500,000 cubic yards of sand moving past some points in a year. The major sand transport occurs between the beach and a water depth of 30 feet. Beyond the thirty-foot contour, the bottom sands are relatively stable and often of different particle size and composition. They are not part of the normal beach sand cycle, except under storm conditions.

Beach sands are composed of varying percentages of quartz sand and shell or coral fragments. Sediments washed down from the mountains of Appalachia are transported to the coast by the rivers of Georgia and South Carolina. From the river mouths they are swept southward by littoral drift along Florida's beaches. As might be expected, the northern beaches have a higher percentage of quartz grains. However, flood control works, reservoirs, and dams constructed on these rivers have impounded the flow of quartz sand. With the source cut off, the present situation on quartz sand beaches is one of net erosion.

The southern and Gulf beaches are composed of less quartz and more shell and coral fragments. The source for this "shell hash" is partly fossil shell and coral eroded from old geologic formations and partly new material originating in the nearshore beds of living shellfish, algae, and coral in water depths of 30 feet or less. When the creatures die, their shells or skeletons are subjected to the grinding and crushing forces of wave action as they are thrown up on shore. These plants and animals are part of a sensitive and complex biological system. Near developed areas they are subject to extreme stress. Pollution from poorly designed sewage outfalls causes disease and ultimate death of the shellfish and corals. Sedimentation

from channel dredging or land clearing can smother them. Increasing turbidity from a variety of sources reduces sunlight available for photosynthesis. The northernmost coral reefs, those nearest heavy development, are dead or dying from an assortment of causes. The net effect, again, is the dwindling or elimination of a source of beach sand.

DYNAMICS. A barrier island or beach is an ever-changing phenomenon. When littoral drift is toward the south, the northern end of the island tends to erode, while deposition takes place at the southern tip. During another season of the year, the processes may be reversed. Under natural conditions, erosion from one beach or bar means deposition elsewhere. As the sea rises at a rate of about 0.03 inches per year, the water depth increases and the shoreline encroaches landward. With increasing depth, wave action increases, and with it, erosion. All these processes are accelerated by storm winds and tides.

A natural beach survives by its flexibility. Like the proverbial reed in the wind, such an area moves with the forces that would destroy a more permanent structure. Unfortunately, when property lines are drawn and coastal areas developed, property owners wish to halt the natural processes which may diminish their holdings. Many engineering solutions have been attempted, but their effect generally has been to increase erosion. Groins and jetties block the littoral drift of sand,

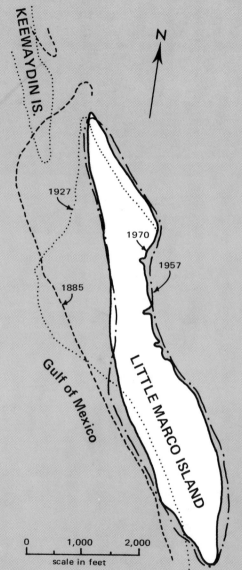

FIGURE 32: Shoreline changes of a barrier island on the southwest coast of Florida. (Source: Lee and Yokel, 1973.)

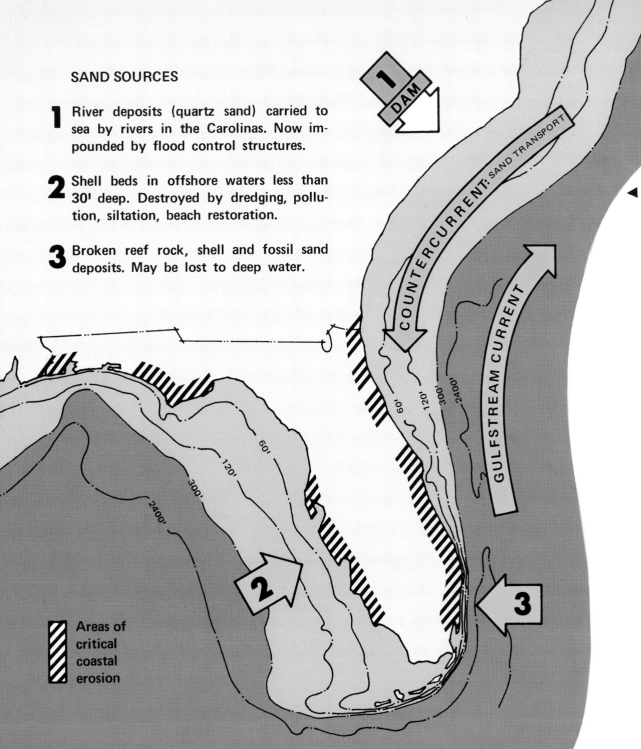

SAND SOURCES

1 River deposits (quartz sand) carried to sea by rivers in the Carolinas. Now impounded by flood control structures.

2 Shell beds in offshore waters less than 30' deep. Destroyed by dredging, pollution, siltation, beach restoration.

3 Broken reef rock, shell and fossil sand deposits. May be lost to deep water.

Areas of critical coastal erosion

DAM

COUNTERCURRENT: SAND TRANSPORT

GULFSTREAM CURRENT

◀ FIGURE 33: Sand is constantly being transported along the shore by currents and countercurrents. When the sources of new sand are destroyed or diminished, the natural cycle of deposition and erosion is unbalanced, resulting in net loss of beach sands. In southern Florida new sand is no longer being transported at former rates, and eroded beaches are no longer replenished naturally.

starving the beaches downstream. These structures temporarily improve the upstream situation, but storm tides eventually carry the excess sand out and over the edge of the continental shelf, where it can no longer be picked up and redeposited by subsequent wave swells. While a naturally sloping beach dissipates wave energy, a vertical seawall reflects it almost completely, creating a scouring action near the toe of the wall. This action causes the undermining and eventual collapse of the wall. The same is true for foundations of buildings constructed too close to the shore. Poorly located channel cuts create similar conditions and can cause erosion of the shoreline and collapse of onshore structures.

The effects of these processes are accelerated on the beaches of the east coast, particularly south of Palm Beach. The bottom slope is much steeper here, and the continental shelf is only two or three miles wide. Even under natural conditions some of the southward flow of sand would result in loss off the edge of the shelf into the depths of the Florida Straits. This would be tolerable if the beaches were still being replenished. With the sources of sand being diminished, however, it is imperative that sound development practices be employed to retain the integrity of the shoreline and the value of beach-front property.

DUNES. Above the berm of the beach beyond the reach of normal wave action, where the sand particles are dry, the wind is the shaping force of dune formation. Dunes are more common on beaches in the northern part of the study area, where more sand is available for their formation. The stabilization process is important, however, even where the dunes are mere undulations in the up-shore sand.

Steady onshore winds pick up dry sand particles and move them inland. Wherever an obstruction to wind flow exists—it can be a clump of grass, a piece of driftwood, or some other object—the wind speed decreases and the sand particles are dropped. If the sand remains there for a short time, and especially if rain consolidates it, fast-growing beach vegetation, such as sea oats or beach grasses, will anchor it in place, creating a larger obstruction where progressively more sand will be deposited. Thus a dune grows. As long as a sand supply is available and dune vegetation keeps the sand stabilized, the dune will remain, or grow, protecting the areas behind it from storm tides, winds, and erosion. The frontal dune is a shifting dune and is extremely fragile. Although it can resist the wind and salt spray, it is vulnerable to vehicular or foot traffic or any other disturbance of its stabilizing vegetation. If the plants cannot keep pace with the moving sand, the shifting dune will be destroy-

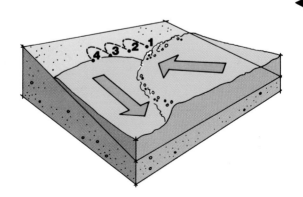

◄ FIGURE 34: Sand transport. When normal waves approach the beach at an angle (arrows) the particles of sand and shell (numbered dots) are transported along the beach in a series of curved movements. Only when attacked by storm waves are the particles moved into deeper water. A subsequent storm may bring the sand back to the beach or deposit it in another place downshore. In this manner wave action constantly reshapens the shoreline and bottom contours.

FIGURE 35: Groins and jetties block the downshore ► drift of beach sands and decrease wave energy. These structures temporarily improve the upstream situation, but beaches downstream are starved and eroded. Eventually, storm tides will carry the trapped sand away from the shore, out and over the continental shelf where it can no longer be picked up and redeposited on shore by subsequent wave swells.

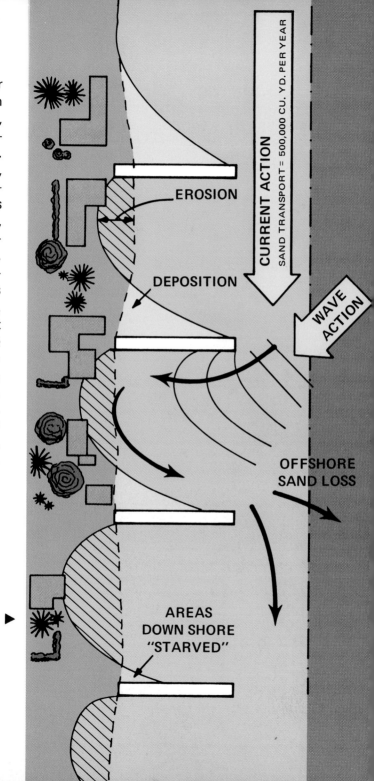

CURRENT ACTION
SAND TRANSPORT = 500,000 CU. YD. PER YEAR

EROSION

DEPOSITION

WAVE ACTION

OFFSHORE SAND LOSS

AREAS DOWN SHORE "STARVED"

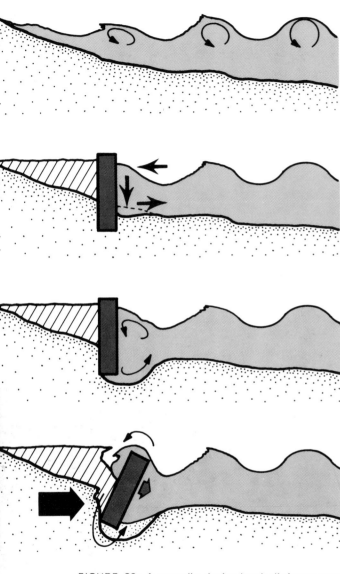

FIGURE 36: A naturally sloping beach dissipates wave energy, but a seawall or building foundation wall reflects the energy almost completely, creating a scouring action near the toe of the wall. This action causes the undermining and eventual collapse of the structure.

ed, leaving the stable dune behind it subject to direct erosion.

Larger plants become established on the stable dune, and eventually the typical West Indies coastal strand forest develops. Sea grape, coco plum, coconut palm, and saw palmetto, along with other shrubs and ground covers, are typical of this plant association. The stable or secondary dune is also fragile, and if the vegetation is removed or damaged by clearing or by dune buggies, pedestrian traffic, fire, or by other means, the dune will then begin to erode, leaving no defense against wind or storm erosion. Behind the stable secondary dune, conditions are favorable for upland vegetation to become established. SCRUB FOREST. The old inland dunes of the coastal margin are stabilized by a different vegetative community than the younger dunes just described. A series of old dune ridges extends from Fort Lauderdale northward through Palm Beach and Martin counties. A few are also found on the west coast between Cape Romano and Charlotte Harbor. They were covered by extensive scrub forests characterized by the sand pine, which has a shorter leaf than the more common slash pine, and by several species of "scrub" oaks, the myrtle oak, scrub live oak, and Chapman's oak. Davis (1943) reports that over 45 plant species may be found in this interesting plant community, making the scrub forest quite variable in number of and relative abundance of species in a given location.

There are a few grasses, and rosemary, saw palmetto, and cactus are common. Most of the vegetation, however, consists of a dense thicket of shrubs and trees.

The sandy soil of the old dunes lacks water-holding capacity; once it is cleared, new vegetation is difficult to establish. A policy of selective clearing and preservation of the existing forest, which is well adapted to these extreme conditions, can save the developer expensive maintenance while providing stabilization of the deep, sandy soil. Unfortunately, early lumbering activity and urban development have eliminated much of this once common plant community.

Beach Restoration. The coastal strand is a region in a state of dynamic equilibrium, both physically and biologically. Development practices that interfere with natural stabilizing processes usually result in the massive erosion and destruction of both beaches and structures. Restoration is extremely costly. Costs have been as high as $1,000,000 per mile, with yearly maintenance costs up to $100,000 per mile. If the conditions that caused the original erosion remain unchanged, restored beaches may be lost again (Monroe, 1975). A 220-foot-wide beach at Fort Pierce, restored in 1971, was reduced to half by 1974.

Restoration operations can destroy sea bottom contours near shore that would have trapped eroded sand until the next storm transported it back to the shore.

Dredging and the resulting siltation associated with both development and restoration projects destroy shell beds, which are a prime source of new sand, and smother bottom vegetation, which stabilizes sediment contours and reduces sand loss. The Crandon Park-Virginia Key restoration project in Dade County has been followed by loss of 50 percent of the near-shore sea grass beds since 1969 (Barada, 1974). Government assistance for restoration is available only if public access to the restored beach is assured. For private owners good development practices are essential to avoid the destruction of property and the astronomical cost of restoration. **Development Practices.** Development strategy in the coastal strand must recognize the dynamics of beach and dune formation and must maximize the natural defenses. Once these defenses are destroyed, man-made defenses must be substituted, and these are expensive and often ineffective.

The beach itself is very tolerant to most recreational use. Foot traffic will not harm it, and it can be safely used for the traditional activities of swimming, sunbathing, fishing, and strolling. It is important to maintain the flexibility of the shore so, if natural erosion occurs, natural deposition can restore the beach without harming structures in the process. The dunes, both primary and secondary, are extremely fragile. The stabilizing vegetation must be maintained since even a small break in plant cover can be quickly enlarged by the wind until a major break is formed. No clearing or major structures should be allowed. Foot traffic is also destructive, so access to the beach should be provided by bridges across the dunes. These bridges can be developed into a design amenity with observation decks and appropriate seating facilities.

Landward of the secondary dune is the most appropriate place for development. If mature trees exist at this point, it is an indication that stabilization has occurred and that the secondary dune will continue to provide protection against ordinary storms and tides. However, unusual storm conditions can still cause destruction from time to time. At best, the coastal strand is a hazardous place for development.

Mangrove

The functions of the mangrove system are discussed in a separate section. In the coastal strand the classic zonation pattern, if it exists, is narrow, and one or more of the typical species may be absent.

Mangroves border the estuaries and lagoons that separate the barrier islands from the mainland. On barrier islands mangroves occur on the inland side in the quieter estuarine waters. Where the island is narrow, coastal strand vegetation may give way to mangroves with no intervening upland vegetation between them. In this situation, very little land is left that is suitable for development. Such locations

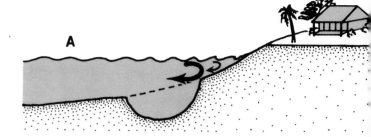
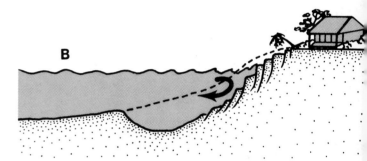

FIGURE 37: Improperly located channel cuts, as in A, can cause slumping and erosion of the shore, as shown in B, with eventual loss of onshore structures. (Adapted from Clark, 1974.)

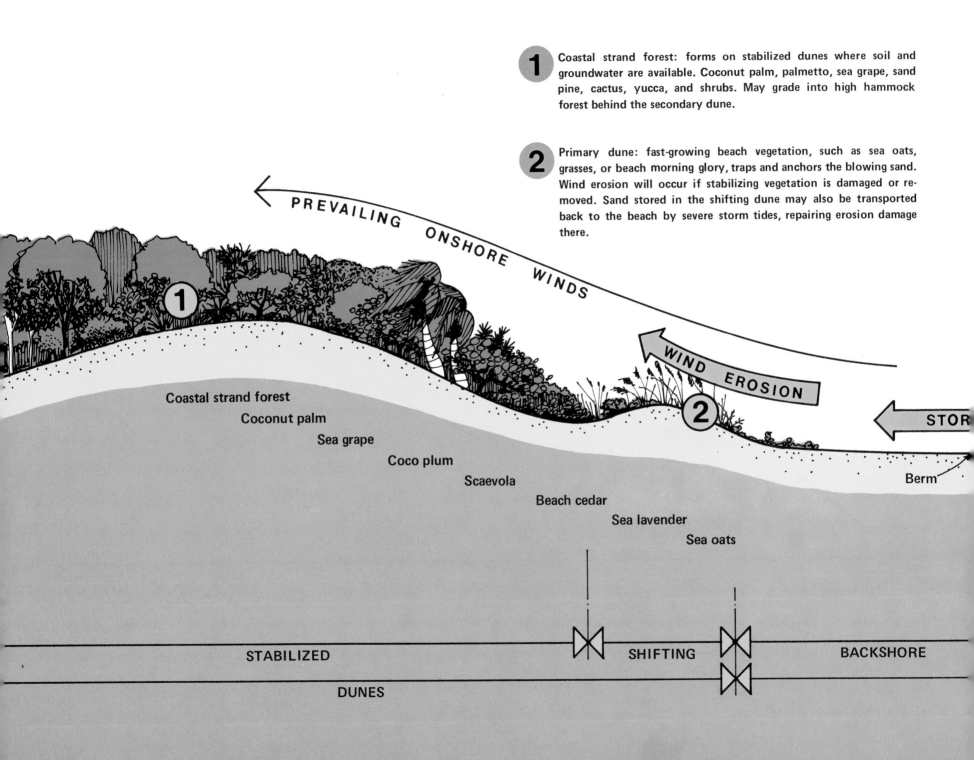

1 Coastal strand forest: forms on stabilized dunes where soil and groundwater are available. Coconut palm, palmetto, sea grape, sand pine, cactus, yucca, and shrubs. May grade into high hammock forest behind the secondary dune.

2 Primary dune: fast-growing beach vegetation, such as sea oats, grasses, or beach morning glory, traps and anchors the blowing sand. Wind erosion will occur if stabilizing vegetation is damaged or removed. Sand stored in the shifting dune may also be transported back to the beach by severe storm tides, repairing erosion damage there.

PREVAILING ONSHORE WINDS

WIND EROSION

STOR

Coastal strand forest
Coconut palm
Sea grape
Coco plum
Scaevola
Beach cedar
Sea lavender
Sea oats

Berm

STABILIZED

SHIFTING

BACKSHORE

DUNES

3 Beach: the beach itself is not sensitive to foot traffic and may be intensively used for recreation without harm. Building on the beach, however, destroys its flexibility and causes erosion.

4 Beach slope: absorbs and disperses wave energy and storm water over a large surface.

5 Littoral drift: sand is transported along the beach by littoral drift. Waves striking the beach at an angle create a current parallel to the shore. If the beach remains flexible and sand sources are not interrupted, over a period of time the amount of sand transported away from a given point will be balanced by new sand transported in.

6 Bottom contours: natural troughs and ridges form traps for sand that is carried away from shore by storm action. Subsequent wave action may redeposit it on the beach.

38 beach and dune

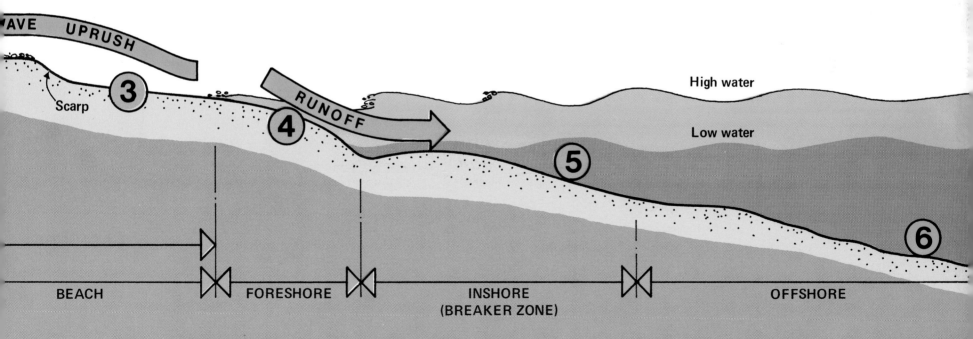

WAVE UPRUSH

3

Scarp

RUNOFF

4

High water

Low water

5

6

BEACH   FORESHORE   INSHORE (BREAKER ZONE)   OFFSHORE

should be eliminated from consideration during the site selection process, since natural constraints and governmental regulations will probably make development difficult, if not unprofitable.

Pineland

On the thin, sandy soils of the rock ridge from Miami southward pines grow slowly, producing a dense and resinous wood, the fabled Dade County pine. Once cured, the wood is so hard that a nail cannot be driven into it. Many of the old pine houses still stand, resistant to termites, hurricanes, and time. Much of the forest, however, is gone. First lumbermen came, then agriculture, and finally urban communities. The rock ridge is well suited for man's use, and the remnants of the pine forests are still disappearing as urban sprawl engulfs them. Although other endangered ecosystems have attracted more attention because they were always relatively rare, pines on the rock ridge may become the rarest sight of all.

The pine-palmetto community, as the

ABOVE. Natural beaches and dunes form a flexible barrier to storm winds and water surge, absorbing their force and protecting upland development. The dune vegetation stabilizes the sand but it is vulnerable to foot traffic and other disturbances.

BELOW. When the dune is built upon, the structures must absorb the full force of storm winds and tides. Erosion caused by this type of development can destroy both dune and beach, leaving the upland unprotected and subject to storm attack.

name suggests, is an open forest of slash pine with an understory of saw palmetto. The ground cover consists of over 100 species of grasses, herbs, and shrubs. This plant community is not tolerant of standing water for more than short periods, and the well-drained ridge provides a suitable habitat. Where there are more humid conditions, the pines give way to hammock forests.

Pineland is maintained by fire. The natural cycle of the ridge included periodic fires that spared the fire-resistant pines and palmettos, but eliminated invading hardwoods. Since hardwoods are shade tolerant but pines are not, hardwoods would eventually shade out and replace the pines were it not for periodic fires. Frequent, less intense fires reduce the amount of dry plant debris, which otherwise would accumulate to a point where it would support a hot fire that could destroy even the pines. The nutrients in the ashes are recycled, and the exposed mineral soil provides the conditions necessary for pine seeds to germinate. Now, however, fires cannot be tolerated near urban development, and hardwoods are gradually invading the pure pine stands.

Agriculture has replaced some former pinelands on the ridge, especially in south Dade County. The surface limestone is crushed by special machinery, and row crops are planted directly in the rocky soil. Tomatoes and beans are two of the main crops shipped to northern markets during the winter. With approximately 500 farms occupying some 80,000 acres in southern Dade County, agriculture plays an important role in the economy of the area. Ninety percent of all vegetables grown are exported to the northern states and account for some $107,500,000 for the recent growing season (U.S. Department of Agriculture, 1975). Dade County is the major producer of winter vegetables for the eastern United States.

The pinelands of the rock ridge present no serious constraints to development. They are well drained, and soil and geological characteristics provide good load-bearing capacity except where solution holes and cavities occur.

The pines themselves are extremely sensitive to injury by construction activities. If the roots are injured by clearing operations, trenching, or heavy equipment, the tree becomes susceptible to the pine bark beetle, which tunnels beneath the bark of the injured tree and ultimately kills it. The process takes several years, and it is common to see pines that were preserved near new housing become yellow and die five or six years later. To prevent this loss, it is necessary to protect the bark and roots from injury during construction. One of the best methods is to leave a wide protective ring of palmettos around the pines. These pine-palmetto "islands," with ground cover or grass planted in the cleared space between them, can be incorporated into an attractive design theme for the landscape. They can also serve as habitat for birds and other small wildlife, a pleasant visual counterpoint to the urban landscape.

Hammock Forest

Hammocks are small hardwood forests, densely vegetated areas that stand out on the horizon as islands of trees. Their outstanding feature is the great diversity of plant species that may be found there. On the Florida mainland, the association is subtropical, with more of the temperate zone species occurring in the northern part of the study area, gradually changing to predominantly tropical species in the south, and becoming purely tropical Caribbean forest on the Florida Keys. About one hundred species of trees and shrubs may be found in the eleven types of hammocks throughout the study area (Davis, 1943). The general features of these types will be described with the ecological regions in which they occur.

The hammocks of the rock ridge are small islands of broad-leaved trees in a sea of pineland. They usually form on a slight elevation under somewhat more humid conditions, perhaps near a sinkhole. As they develop, the dense growth tends to increase humidity within, creating a microclimate that is warmer in winter and cooler in summer than the surrounding area (Craighead, 1971). Partially decayed leaf litter fills the solution holes and builds up on the hammock floor to eventually form a loamy, moisture-holding peat soil a foot or more deep. The combination of moist soil and humid air keeps out most pineland fires. Craighead (1971) points out that when fire does destroy the hammock some roots survive, protected in the solution holes of the rocky substrate. New growth sprouts from these roots and another forest develops, almost identical to the old. Repeated burning, however, will result in permanent destruction. Today, in the absence of the natural fire cycle on the ridge, hardwood hammock species are expanding into the pinelands. Alexander (1973) has observed that the transition from mature pineland to hammock can occur in twenty-five years. On the exterior, the typical hammock is edged with a thicket of shrubby growth or a dense band of saw palmetto that appears to be impenetrable. This curtain of vegetation keeps breezes out and the interior humidity high, and sometimes serves as a firebreak. This edge is also the transition zone where pioneer species begin to invade the surrounding pineland. Within the hammock, however, the understory growth opens up beneath the tall canopy. Rather than a jungle, a mature hammock resembles a rain forest.

The dense growth of the subtropical hammock includes up to fifty species of trees and shrubs. Large tropical trees forming the forest canopy may be strangler fig, wild tamarind, gumbo limbo, mastic, bay, pigeon plum, several kinds of

FIGURE 39: Design concept for bridging the dunes.

stoppers, and rarely, the royal palm and mahogany. Among the typical trees of the temperate zone are several kinds of oak, mulberry, and occasionally red maple. Beneath this high canopy, such smaller trees as paradise tree, poisonwood, satinleaf, and lancewood fill the intermediate space. Shrubs include wild coffee, myrsine, and marlberry from the tropics as well as such temperate zone species as beauty berry, hamelia, southern sumac, and wild lime. On deeper, sandy soils oaks may dominate the hammock association, forming large stands with cabbage palms and saw palmetto.

The even temperature and high humidity of the hammock microclimate provide an ideal habitat for a diverse assortment of interesting smaller plants. Where light is sufficient, vast fern gardens develop, containing both common and rare species. In the canopy the branches of some trees, covered with vines and dense epiphytic growth, resemble hanging gardens. These plants are not parasitic, but merely use the host plant for support. The epiphytes of the hammock include ferns, bromeliads or airplants, and orchids. Some of the species are rare, and some have been so enthusiastically collected that they may have disappeared altogether (Craighead, 1971; Davis, 1943).

Once over 500 hammocks dotted the pineland ridge of Dade County alone, but most have been eliminated in the last thirty to fifty years (Craighead, 1971). Some were in the path of urban development, others were eliminated by fire after extensive drainage lowered the water table and reduced soil moisture and humidity. Although hardwood species are encroaching upon the pinelands, it will be years, if ever, before natural succession results in formation of a mature hammock forest. Remaining hammocks are an environmental amenity that should be preserved.

Where possible, the hammocks should be left intact, whether in private or public ownership.

Hammocks contained within large development tracts are a valuable feature that should be retained and utilized. Access can be created so that the hammock can be used as a private park for the residents of the new community. Where such preservation is not possible, or where damage has already occurred, buildings should be clustered and other construction confined to disturbed areas as much as possible to preserve the viable portions of forest. Roads and parking should be kept to a minimum, with roadways oriented to act as breezeways. Removal of the perimeter shrub thicket will also increase air circulation and reduce temperature and humidity. Proper site planning could probably save 50 percent of the hammock growth.

84

summary

Environmental Services

Beach and Dune
Recreational resource and tourist attraction

Natural shoreline protection. Beach face absorbs and dissipates energy of storm wave uprush. Dunes protect the upland.

Mangroves
Stabilize shoreline

Protect upland development by absorbing energy of wave surges

Build land by accumulating sediments and other materials

Improve near-shore water quality by filtering storm water runoff

Provide habitat and food for sport and commercial fish

Marine Grass Beds
Provide habitat and food for sport and commercial fish

Stabilize shallow water bottom contours

Act as sediment and nutrient traps

Hardwood Hammocks and Pine Forest
Aesthetic value

Provide cooling by evaporation and by shading

Stabilize soil, prevent sedimentation from runoff

Increase value of the site if preserved

Require less maintenance than other landscaping materials

Provide habitat for wildlife

Constraints

Generally, the coastal ridge is suitable for development without excessive constraints. Caution is needed in wetlands and the coastal fringe.

Destruction of natural beach and dune can result in loss of beach to storm action. Restoration costs are extremely high. Breakwaters, jetties, and groins are seldom successful.

The coastal construction setback line has been set by the state at fifty feet inland from mean high water, an interim regulation, to be superceded by a setback line based on topographic, vegetative, and dynamic criteria for each county.

Development should be concentrated on the ridge, where soil suitability is high and natural drainage is adequate. Marginal lands should be developed with extreme caution, if at all, to prevent environmental damage and to avoid future expenditures for flood and wind damage.

Opportunities

Shorefront construction should be kept behind the stable dunes.

Foot traffic and other uses that would destroy stabilizing vegetation should be prohibited on primary dunes.

Intensive recreational use of beach is not harmful. No construction on the beach should be permitted.

Mangrove areas that border coastal waters or tidal channels and estuaries should be preserved as buffer zones to maintain water quality and shoreline protection.

Residential construction in the hurricane flood zone should have a first-floor level above the 100-year flood elevation. This can be accomplished by "stilt" construction. Filling may occur in upland areas.

Natural vegetation should be preserved wherever possible to minimize erosion, retard runoff, and maintain aesthetic values and habitat.

Landscaping should utilize native species, since they require less maintenance, fertilizer, and water.

Urban runoff should be directed to vegetated swales and holding ponds, rather than to curbs, gutters, and storm drains. This allows natural processes to improve water quality and allows time for aquifer recharge to occur.

Sewage treatment plants should be located in areas not subject to flooding. Septic tanks are not suitable in such areas.

Bulkheading should be avoided where possible, and natural systems, such as mangroves, should be preserved instead. Sloping riprap is preferable to vertical bulkheading.

the florida keys

The Florida Keys are a uniquely beautiful chain of 97 islands, 35 of them linked to the peninsula by the Overseas Highway. The word "Key" originated as the Spanish "cayo," meaning a low, small island. These islands are covered by tropical hammock or pine forests and are fringed by mangroves. The clear blue waters contain coral reefs and provide a habitat teeming with marine life. The environmental system of the Keys is extremely fragile, and the complex interrelationships of its subsystems are only beginning to be understood. Development here must be sensitive to special problems if the environmentally and economically important resources are to be maintained.

Geology

As in the rest of Florida, limestone underlies the Keys. The upper Keys, north of Big Pine Key, are ancient coral reefs that were exposed when sea levels gradually fell. The lower Keys are composed of oolitic limestone, formed by precipitation of tiny calcium carbonate particles from seawater. These subsequently were cemented together to form oolite. The porous nature of these limestones allows relatively free movement of groundwater. Any rainwater that may percolate into the aquifer is soon mixed with intruding seawater, creating a shallow, brackish aquifer that is unsuitable as a potable water supply. The Floridan Aquifer underlies the Keys at greater depths. Although it contains brackish water, the cost of extraction and treatment would be lower than for the desalinization of seawater.

The permeability of these formations and the high water table have important implications for both solid and liquid waste disposal. Methods commonly used elsewhere cannot be used in the Keys because the effluents quickly migrate through the porous rock into near-shore surface water, degrading water quality. Proposals for deep well disposal of sewage waste into the boulder zone of the Floridan Aquifer should be weighed against the aquifer's potential as a water resource.

Topography

The highest elevations in the Keys are found in Key Largo; there a central ridge reaches 12 to 15 feet. In Key West about 250 acres are 8 to 15 feet above sea level. Ninety percent of the land area in the Keys, however, is less than 5 feet in elevation. The Keys lie on a shelf that extends westward beneath Florida Bay and eastward to the fringing reefs. The shallow waters barely cover mud banks and patch reefs. Beyond the outer reefs to the east, the bottom slopes rapidly to the depths of the Florida Straits. This combination of low elevation and exposure to shallow water on all sides make the Keys extremely prone to flooding by storm-elevated sea levels.

Hurricane Hazard

The hundred-year storm (a 1 percent chance of occurrence in any given year) would produce a potential 12-foot storm surge in the upper Keys and a 7-foot 6-inch surge in the lower Keys. This potential, however, is not the greatest surge that may be experienced. The 1935 hurricane produced an 18-foot surge at Lower Matecumbe Key, drowning 400 persons. Tidal surges are more damaging to coastal areas than hurricane winds (Florida Department of Natural Resources, 1974, p. 7).

With the exception of high areas on Key Largo and Key West, all the Keys are below the statistical 100-year flood elevation. This low elevation poses serious implications for development. Careful assessment of potential dangers to life and property can result in development practices that minimize those dangers. Development strategy should include two factors. First, existing natural protection should be maintained. The barrier reefs are a natural breakwater that can lessen wave force. The mangrove fringe is a second line of defense that can absorb considerable wave energy if the mangroves are dense and healthy and if they occupy a sufficiently wide band between the shore and developed upland. In addition, the

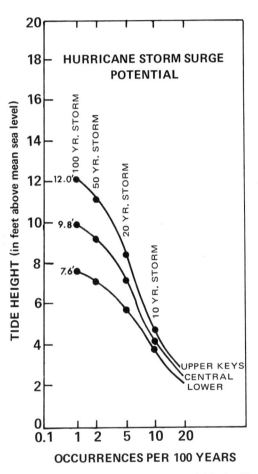

FIGURE 40: Hurricane storm surge potential in the Florida Keys. (Source: Florida Department of Natural Resources, Coastal Coordinating Council.)

mangrove root structure stabilizes the shoreline and helps prevent erosion.

Second, adequate building design practices can also help to minimize damage. The Federal Flood Insurance Law requires such practices; the first floor level must be above the 100-year flood elevation. In addition, lower portions of buildings should be left open to minimize resistance and allow flood waters and debris to pass freely through. This is advisable not only for the physical safety of the occupants, but also because construction designed to resist flood forces is more costly to build.

Vegetation

The scenic value of Keys vegetation is one of the great natural assets of the area. Aside from producing environmental damage, the practice of totally clearing building sites creates ugly areas of glaring white, exposed limestone. The subsequent attempts to landscape these areas are difficult and costly because most commercial available landscape materials are poorly adapted to the rigorous conditions of constant salt-laden sea breezes and seasonal lack of rain. Such landscaping requires expensive inputs in the form of topsoil, freshwater, fertilizers, and labor. Maintenance costs will far exceed initial cost. The replacement value of the native vegetation removed would probably astound the developer. Particularly in the tropical hammock, selective clearing can save many thousands of dollars worth of valu-

41
the florida keys

1 Bottom muds: easily disturbed, fine-textured muds, composed mainly of the remains of algae (*Penicillus* and *Halimeda* spp.). Hard bottom is found where current action scours away bottom deposits.

2 Shallow water of Florida Bay: highly productive, especially for shrimp. Sensitive to turbidity produced by dredge and fill and to pollutants from sewage and urban runoff.

3 Tidal zone and near-shore waters: nursery for juvenile stages of fish and shellfish.

4 Mangroves: produce food for many marine species, stabilize the shoreline, protect the upland from storm damage.

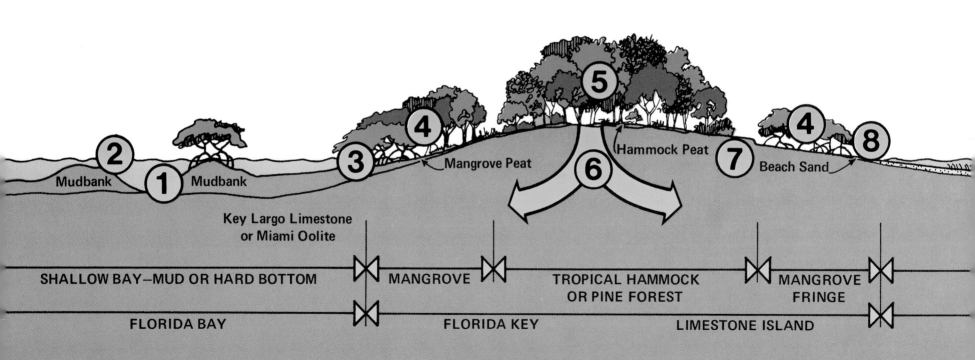

Mudbank — 2 — 1 — Mudbank

3 — 4 — Mangrove Peat — 5 — Hammock Peat — 7 — Beach Sand — 4 — 8 — 6

Key Largo Limestone
or Miami Oolite

SHALLOW BAY—MUD OR HARD BOTTOM ⋈ MANGROVE ⋈ TROPICAL HAMMOCK OR PINE FOREST ⋈ MANGROVE FRINGE ⋈

FLORIDA BAY ⋈ FLORIDA KEY ⋈ LIMESTONE ISLAND ⋈

SOURCES: Enos, 1971; Fla. Div. of State Planning, 1974;
Fla. Dept. of Natural Resources, 1973a.

5 Upland forest: tropical hammock in the upper Keys. Typical West Indies forest, unique and valuable, containing tamarind, pigeon plum, poisonwood, Jamaica fish poison, gumbo limbo, and many other tropical species. In the lower Keys, the upland is covered with pines, thatch palm, palmetto, grasses, and shrubs, similar to the pine forest of the coastal ridge.

6 Porous limestone: allows rapid mixing of fresh rainfall with seawater and rapid migration of sewage pollutants from septic tanks and injection wells through rock to near-shore waters.

7 Escarpment: low, wave-cut rock ledges, formed at former sea levels. Usually occur at the transition of mangrove and high hammock. Little vegetation—good location for road.

8 Beach: narrow sand beach on some keys, rock on others. Extreme caution is necessary to prevent loss of sand by erosion.

9 Grass beds and bottom-anchored algae: act as sediment traps, retard erosion, stabilize the bottom. Grasses provide nutrient uptake, enhancing water quality. Algal skeletal structures break down to form beach sand. Grass beds are damaged by urban runoff, siltation from upland development.

10 Sand shoals: current action sweeps sand from bottom in some areas, depositing it in quieter waters to form banks and shoals.

11 Patch reefs: occur in the "shadow" of a key (island). They are absent opposite channels because they need protection from the turbid, sediment-laden water transported by currents from Florida Bay. When new channels are opened through, patch reefs are destroyed.

12 Barrier (fringing) reefs: like patch reefs, these are built by coral polyps in clear tropical waters. These highly productive biological systems also provide protection from storm waves. High economic value for tourism, sport fishing.

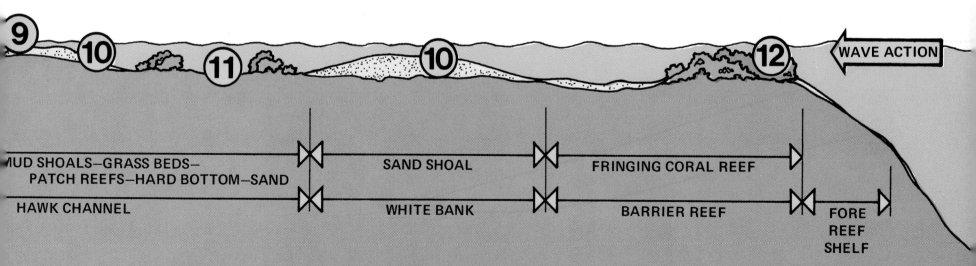

able trees. Moreover, since these trees are already established and adapted to climatic conditions in the Keys, they require little maintenance.

Mangroves. The mangrove system is described fully in another section of this book. Its function in storm and erosion protection is particularly valuable in the exposed shorelines of the Keys. Mangroves also are vital as nursery grounds for both sport and commercial fisheries, major components of the Keys economy.

Reef System

The reefs of southeast Florida are composed of tropical corals occupying the continental shelf from Palm Beach county southward to Key West and the Tortugas. They lie to the east of the Florida peninsula and Keys, with the fringing reefs at the outer shelf edge and patch reefs between them and the land mass. In the days of sailing ships, the reefs claimed many victims, giving rise to the time-honored Keys profession of "wrecking." When a ship foundered on the reefs, whether driven there by storm or lured there by a mysteriously misplaced navigation light, the wreckers would row out to salvage what they could. They lived a comfortable life on the proceeds. Today, the reefs are well marked, and their fascination lies in the exotic beauty hidden beneath the surface of the sea.

Living coral reefs are an extremely important part of the biological cycle of marine systems. The reef itself is built by colonies of small, flowerlike animals called coral polyps that secrete the calcium carbonate (limestone) which forms the solid structure of the reef. The polyps live in association with algae, and their mutually beneficial relation is necessary to the survival of each. The cycle is similar to the relation of plants and animals on land. The algae produce oxygen and food materials that are utilized by the polyps. In turn, the algae thrive on the waste products of the polyps. Both depend on nutrients from the surrounding waters. Also necessary for their survival are clear, shallow (less than 150 feet) water, to allow the penetration of sunlight, and a fairly stable water temperature above 70° F but no more than 90° F.

The southern Florida reefs are at the northernmost limits of their temperature range, and thus are under considerable natural stress. When additionally stressed by human activity, they may be unable to maintain themselves. Much damage is done to the reef by collectors of coral or tropical fish. Blasting has been used to break up the reef, along with less drastic methods, such as pry bars or other hand tools. Boat anchors are also extremely destructive.

Aside from such structural damage, there are extensive areas where the coral polyps are dying. The cause is unknown. Some authorities point to the siltation resulting from near-shore and upland dredging as the most important cause of coral reef decline. Aside from physical smothering, turbid water also stops the penetration of sunlight necessary for photosynthesis. Winter storms and winds also stir up bottom sediments, which can stay in suspension, turning the water milky for days. The water clears between storms, however, allowing the corals to recover. Dredging is continuous, on the other hand, and does not allow this recovery period. Biologists cannot as yet determine a specific cause of coral decline although they have investigated and recorded some problem areas (Voss, 1973.) The fringing, or barrier, reefs in clear water are in better condition than the inner patch reefs, the latter being closer to coastal activities that tend to degrade water quality. Reefs in the lower Keys are healthier than those in the upper Keys (Voss, 1973). Hen and Chickens Reef, off Key Largo, is estimated to be 85 percent dead (Florida Department of Natural Resources, 1974). The upper Keys have been the scene of unprecedented development activity in recent years, much of it of the dredge-and-fill type. Almost all residential units are serviced by septic tanks. The upper Keys are also more affected by southward-flowing currents that carry sewage as well as urban and agricultural runoff from Dade County.

Apparently there may be no single cause of reef mortality, but rather a lethal combination of natural stress, declining water quality, and physical damage.

Coral reefs serve several important functions in the southern Florida system. They are among the most productive biological systems, not because of the corals themselves, but because they provide a habitat for other life forms. Many species of sport and commercial fish depend on the reef. Along with estuarine and mangrove habitats, the reefs produce a significant part of the total value of the fisheries industry. In Monroe county, alone, the total market value of all fishery products in 1973 was estimated to be between $70 and $90 million. In addition, the value of sport fishing to the Keys economy is estimated at $25 million annually (Florida Department of Administration, 1974, p. 25).

Aside from sport fishermen, the reef attracts other tourists who come to view and photograph the only living coral reef in the continental United States. In 1973 John Pennekamp Coral Reef State Park alone attracted 376,000 visitors (Florida Department of Administration, 1974, p. 11). Estimates of the economic impact of these and other reef visitors are not available, but tourism would undoubtedly be negatively affected if the reefs were destroyed.

The reef also functions as a natural breakwater. Both normal currents and tropical storms generate erosive forces that are considerably reduced as the waves pass over the reef before striking the shore. Hurricanes can break off great coral heads, but the living reef can repair itself in a few years (Voss, 1973). Death of the corals would mean that the reef itself would begin to disintegrate and erode. New growth would no longer occur to repair damage, nor would reef building keep pace with the rising sea level. Wave and storm forces would then have a greater impact on the shore, eroding waterfront property and increasing the cost of shoreline protection (Florida Department of Administration, 1974).

summary

Environmental Services

Reefs
  Recreation, tourist attraction
  Habitat and food for sport and commercial fish
  Shoreline protection by buffering wave action and storm surge
Mangroves
  Stabilize shoreline
  Protect upland development by absorbing energy of wave surges
  Build land by accumulating sediments and other materials
  Improve near-shore water quality by filtering storm water runoff
  Provide habitat and food for sport and commercial fish
Marine Grass Beds
  Provide habitat and food for sport and commercial fish
  Stabilize shallow water bottom contours
  Act as sediment and nutrient traps
Beaches
  Recreation resource and tourist attraction
  Natural shoreline protection and buffering
Hardwood Hammocks and Pine Forest
  Aesthetic value
  Provide cooling by evaporation and by shading

Stabilize soil, prevent sedimentation from runoff

Increase value of the site, if preserved

Require less maintenance than other landscaping materials

Provide habitat for wildlife

Constraints

Lack of freshwater, except by aqueduct from the mainland, will be a limiting factor if the water supply cannot be expanded. Desalinization is costly.

Porous geologic structure and high water table permit effluent from septic tanks or seepage from solid-waste disposal sites to pollute near-shore waters, degrade recreation value, destroy reefs and other marine life. Waste disposal problems are a limiting factor that will require technologic solutions.

Linear configuration of the Keys makes central services such as water or waste treatment difficult to plan and costly to operate.

Hurricane and flood hazard potential is high, since 99.6 percent of Keys land area is below the 100-year storm level. Evacuation is impractical, since U.S. 1 is the only road to the mainland.

Sedimentation from runoff during construction activity is detrimental to water quality and marine life. Preventative measures should be used during construction, and cleared areas should be promptly replanted.

Opportunities

Domestic freshwater supply could be augmented by cisterns to collect and store rainwater. This is a common practice in the Caribbean.

Water-conserving plumbing fixtures are available and should be required in new construction or renovation.

Private reverse-osmosis treatment plants, using water from the Floridan Aquifer, may be a solution for providing freshwater to large developments.

Waste treatment is a serious problem, and a combination of solutions is necessary. Innovation and cooperation between government and developers in finding new techniques is essential. Pace of development should be limited to the capacity of available services.

Water-oriented transportation, such as high-speed passenger ferries and barges for movement of goods, could reduce the need for extensive highway expansion, which degrades visual quality and consumes valuable land at the expense of other land uses.

Development planning should include selective clearing and clustering of built units to maximize preservation of natural vegetation and ground covers. Cleared areas should be planted as soon as possible to prevent runoff and sedimentation. Native plants species should be used.

On-site retention of runoff must be provided, as discussed in Part Three.

Planned unit developments, clustered to minimize site clearing, should be encouraged. Bonus densities could be given for site preservation and setback from the water's edge.

Cluster developments with common marina facilities are preferable to conventional canal developments.

Bulkheading should be avoided where possible, and natural systems, such as mangroves, should be preserved instead. Sloping riprap is preferable to vertical bulkheading.

the sandy flatlands

The low, pine flatwoods of the sandy flatlands cover a large part of the Florida Peninsula and the coastal plain of Georgia and South Carolina. In southern Florida the western flatlands extend north from the Big Cypress watershed and from the coastal strip to Lake Okeechobee. On the east the eastern flatlands extend from the shore of the lake and the eastern edge of the Everglades to the Atlantic coastal ridge.

Topography

The flatlands, as the name implies, are relatively level remains of old sea terraces. The nearly flat or gently rolling land surfaces are dotted with small ponds and sloughs. Fossil dunes and sand ridges delineate the old shorelines. The role of the sea in shaping this landscape can easily be seen. The western flatlands attain a height of over fifty feet near the northern boundary of Charlotte and Glades counties—the highest elevation in the study area. The maximum elevation in the eastern flatlands is just over thirty feet in Martin County, north of the St. Lucie Canal.

Geology and Soils

The sandy terrace deposits overlie relatively impermeable marl formations. The shallow soils drain rapidly, until rainfall raises the shallow water table and water reaches the surface in the numerous shallow depressions. The depressions form ponds and drainageways that are permanently wet in areas of low elevation, but may only hold water during the wet season at higher elevations. At varying depths below the surface the sands often contain a cemented layer of hardpan. This layer may be permeable or it may inhibit percolation and create conditions of standing water.

The Eastern Flatlands

The flatlands on the east side of Lake Okeechobee extend southward to Loxahatchee, according to Davis (1943). Parker et al. (1955) describe them as extending to Coral Gables. The difference may be explained by calling this southern strip a transition zone dissected by many wetlands and drainageways (transverse glades) between the Everglades and the coast. The northern portion is more typical of true flatlands.

The northern part of the region is drained by the Allapattah marsh, a long, shallow slough crossing Martin and St.

Lucie counties. To the south it splits into two forks, one discharging into Lake Okeechobee and another draining into the Everglades. The Loxahatchee and Hongryland sloughs lie to the south in Palm Beach County. The St. Lucie Canal crosses the flatlands from Lake Okeechobee to the Atlantic. There are many narrow sloughs and seasonally wet ponds that range in depth from 1 to 4 feet.

South of the Loxahatchee slough the flatlands extend past Coral Gables, where they abut the limestone ridge and are covered by Everglades soils (Parker et al., 1955).

Western Flatlands

The western flatlands grade into the Big Cypress on the south and the coastal strip on the west. The boundaries between the regions are not well defined.

The Caloosahatchee River and its valley divide the region. In the southern section the high elevations near the town of Immokalee form a drainage divide, and surface water flows northward to the Caloosahatchee and southward to the Big Cypress. Elevations are generally less than thirty feet, lower than the northern section. Thin, sand deposits overlie marls or limestone, making the soils alkaline or neutral, a condition favorable to the formation of cabbage palm hammocks. Marshes, swamps, and wetlands are numerous. Corkscrew Swamp, Okaloacoochee slough, and Twelve-mile slough are similar to the wetlands of the Big Cypress Swamp.

North of the Caloosahatchee River the flatlands are higher; consequently, some parts are fairly well drained by streams, although many ponds and water-filled depressions remain. Larger areas of unforested palmetto prairie are now being extensively used for pasture.

Sandy Wetlands

In the transition area between the drier flatlands and the adjoining wetlands of the Everglades and Big Cypress, the soils are sandy with some muck and marl intermixed in pockets. The difference between the sandy wetlands and the higher flatlands is a high water table; surface water often stands throughout the year. It is probable that the sandy wetlands of the eastern flatlands serve to recharge the aquifer in areas where they overlie permeable geologic formations, especially the Anastasia formation.

Vegetation

Pine Flatwoods and Dry Prairies. The nearly level sandy pine flatwoods are covered with open pine forest and grasses; saw palmetto is the most prevalent shrub. Similar grasslands with saw palmetto,

but without pines, are called palmetto prairies or dry prairies. In the flatwoods the pines are so widely spaced that they are of secondary importance, and these forests look more like grasslands. The trees are usually tall, with few lower branches, and they grow slowly in the thin soil. Many of the forests have been cut over and are now saw palmetto prairies. Absent in lower wet soils, the saw palmetto grows tall and forms dense thickets on thin sands. Shrubs of the heath family grow where the soil is acid. Common in the low pinelands are gallberry, running oak, and pennyroyal mint. Wildflowers and sedges abound in both flatwoods and prairies.

Toward the south, on thinner alkaline soil over limestone or marl, cabbage palms and wax myrtle become more abundant (Davis, 1943).

Hammocks　　Hammock formation is dis-
cussed in the section describing the coastal strip. The hammocks of the flatwoods contain a greater percentage of temperate zone species. They are often dominated by the live oak in association with cabbage palms. Other hammocks known as "cabbage woods," are almost pure stands of cabbage palm.

Ponds and Wetlands

One of the most distinctive features of the flatwoods and palmetto prairies are the many shallow ponds and drainageways. The fluctuating depths of water cause a zonation of plant associations to occur in these depressions. The deep center may contain aquatic plants that require standing water, bordered by a zone of such marsh plants as pickerelweed, arrowhead, fire flag, and saw grass. The width of these innermost zones depends on the angle of slope of the depression
and the seasonal depth of water. The outermost zone is wide and dominated by a wet prairie vegetation of grasses and sedges. The zonation is variable, depending on local conditions (Davis, 1943).

Over most of the sandy wetlands the dominant plant community consists of wet to dry prairie, with a mixed association of grasses, sedges, and other herbaceous plants, but very few trees. In Palm Beach and northern Broward counties, the sandy wetlands are occupied mostly by open cypress forest, interspersed with prairie vegetation.

Freshwater swamp forests also occur as a subregion where the land is low and flooded most of the year. Plant species found here include cypress, water oaks, willow, bay, holly, and abundant freshwater aquatics. These wetland communities are described more fully in the sections describing the Everglades and Big Cypress Swamp.

summary

Constraints

When drainage is adequate there are few constraints to development. When the water table is high, however, flooding may occur during part of the year.

An impermeable cemented layer of hardpan may occur at varying depths beneath the surface. This layer may cause localized ponds to develop and may present special problems for foundations.

"Cap" rock may occur extensively at the surface, presenting special design problems, depending on the depth and bearing capacity of the rock.

In some areas, particularly in the eastern flatlands, extensive wetlands serve as municipal water supplies. In these areas lowering of the water table by drainage for development should be undertaken only in the context of a regional water management plan.

Opportunities

The sandy flatlands provide good opportunities for development where the water table is not high. The design concept for the Port La Belle community in Glades and Hendry counties is an example of good development practices fitted to a suitable site. Site constraints were considered carefully and became amenity features.

The design concept for the development was based on a drainage system that imitates the natural system.

The basic feature is the use of interconnecting ponds and water courses incorporated into a greenbelt network. This network brings areas of natural beauty to all parts of the development, while providing an effective storm drainage system. Greenbelt drainage areas range in width from 20 feet at back lot lines to 50 to 275 feet for water course right-of-way.

Runoff is directed from street swales or overland flow into a secondary drainage system of wide, shallow swales about 3 feet deep, with a side slope less than 6:1. The swales are grassed and dry most of the time. Overgrowth or debris can be easily removed.

The primary system is an extension of the canals and natural drainageways that originally existed on the site. The natural sloughs are to be kept intact, and natural watercourses are to be incorporated into the total development concept. The system consists of grassed swales connected with artificial ponds, which serve as borrow pits, storm water storage reservoirs, ecological controls, and visual amenities. Ponds will normally contain 5 to 6 feet of water.

Control structures will be used to regulate water flow. The objective is to maintain a smaller rate of flow over a longer period of time. Discharge of runoff to the Caloosahatchee River is reduced, and water quality is enhanced by allowing time for debris to settle out, vegetative uptake of nutrients to occur, and seepage to replenish the aquifer and maintain the water table.

Native trees, such as oaks, will be planted along the greenbelt drainageways to help absorb nutrients. Where large existing trees occur, the street swales and drainageways are either diverted around the trees or have a steeper slope. The minimum distance from the tree trunk is 8 feet. Smaller trees are to be transplanted to other locations.

the interior wetlands

The interior wetlands of southern Florida—the Everglades and the Big Cypress Swamp—are a single ecological unit, the water resource that supports human and animal life of the region. To understand how this system functions and the problems associated with its management, the components of the aquifer, that is, geology and water, should be considered together.

Groundwater and the Aquifers

Groundwater occurs beneath the surface of the earth in the zone of saturation, where it fills all the voids, fissures, solution holes, or other openings. Water that infiltrates directly into the ground and fills the voids between the particles creates a zone of saturation that is in contact with the atmosphere at its upper surface. This surface is called the water table. The elevation of the water table generally rises as the land surface rises, and, in general, it reflects the topography of the land surface. Where the water table is higher than the land surface, lakes, streams, or marshes occur. The geologic formation in which water is held, and from which water can be collected, is called an aquifer.

The Artesian Aquifer. When water fills a permeable layer of rock that is confined between impermeable layers, this water is under pressure, both from water entering at a higher level where the aquifer lies at the surface and from the weight of the overlying geologic formations. Wells drilled into such an aquifer relieve the pressure, and water rises in the well without pumping. Such a well is called an artesian well. The area where water enters the aquifer is called the recharge area, because it is there that water originally enters the aquifer.

The Nonartesian Aquifer. This type of aquifer contains water that is not confined at its upper surface by an impermeable layer, but instead is in contact with the air at the water table. Unlike the artesian aquifer this type of aquifer is not recharged in a remote place, but receives its recharge waters directly from percolation over its entire surface, whether from rainfall or from overland sheet flow.

Both types of aquifer occur in southern Florida. The Floridan Aquifer is artesian and underlies all of the study area at great depth; it is about 900 feet below sea level at Miami. Its recharge area is in central Florida; by the time the water reaches southern Florida, it has passed through miles of limestone, some of which still

contains chlorides (salt) in pockets left behind after the sea receded. Wells drilled to the Floridan Aquifer yield brackish water containing dissolved calcium carbonate (limestone), making it too saline and mineralized to serve as a potable water supply for southern Florida. Parker (1955) describes it as saline, sulfurous, and corrosive. Old, abandoned artesian wells can contaminate the shallow aquifer above with saline water. Parker (1955) reported that an abandoned oil exploratory well near Pinecrest in Monroe County, after leaking for years through its corroded casing, had killed vegetation over a radius of 100 feet and had contaminated groundwater over a much greater distance. The Floridan Aquifer yields good water, however, in the central and northern part of the state.

The shallow, nonartesian aquifers supply potable water in southern Florida in quantities and qualities that depend on several factors. First is the water-bearing capacity of the geologic formation itself. Some formations, such as the Anastasia and Miami oolite of the coastal ridge and the Key Largo limestone of the Keys, are very permeable, allowing water to enter freely. They lack holding capacity, however, so water drains out easily when not confined, as in canal cuts or at the coast. Salt water can also enter easily, and in the Keys water rises and falls in wells in response to the ocean tides (Parker, 1955). Some formations are dense and have very

little space between the rock particles and therefore are relatively impermeable. If water is pumped from a well in such a formation, water cannot flow fast enough through the rock to replace what was pumped out and the well may soon run dry.

The best shallow aquifers in southern Florida are those that occur in limestone that is riddled with solution cavities. Slightly acid groundwater dissolves away the rock through which it percolates, and such cavities become partly filled with sand. This structure enables the formation to hold enormous quantities of water and to transmit it rapidly to wells. The Biscayne Aquifer of Dade and Broward counties is of this type.

Quality of water contained in the shallow aquifer can easily be affected, since any contaminants that can reach the water table can also affect the aquifer. These include surface water pollutants in lakes, canals, and wetlands, as well as direct percolation of contaminated water through overlying rock or soil. Some pollutants are eliminated by mechanically filtering water through rock and by the active chemical and biological processes that occur as water passes over or through soil and vegetation. Other contaminants, such as heavy metals contained in agricultural chemicals or runoff from highways and parking lots, are more persistent and can remain in the water indefinitely. Organic pollutants, which ordinarily can be processed by soil

Oil well

or vegetation, can degrade water quality if they are too concentrated. For example, septic tanks have been used in the region for many years, with no harmful effect to the water supply until increasing population density created a volume of organic waste too large and concentrated for the natural system to process. The symptoms of overload are foul-smelling, algae-filled canals and domestic wells contaminated with coliform bacteria.

Another potential contaminant is seawater. It is the pressure of freshwater that

Solution cavity in limestone

Sanitary landfills can contaminate the shallow aquifer

prevents seawater from intruding into the permeable limestone of the shallow aquifer. First, since freshwater is less dense than seawater, it is lighter and tends to "float" above it in a lens-shaped layer, unless mixing occurs. Second, the interior wetlands are at an elevation, however small, above sea level. This elevation creates a pressure gradient and causes the freshwater to flow seaward through the porous rock. The salt front, where freshwater meets salt water in the aquifer, is located somewhere near the coast, de-

pending on how much freshwater pressure is exerted (also see Figures 16 and 17).

If the level of freshwater in the aquifer drops to nearly sea level, salt water can intrude easily. The deeper the aquifer, the greater the difference must be between freshwater levels and sea level. This difference is called the freshwater head. Theoretically, according to the Ghyben-Herzberg principle, a 1-foot head of freshwater will depress salt water to 40 feet below sea level. The Biscayne Aquifer, for example, is approximately 100 feet deep in

Dade County and would therefore require a 2.5-foot freshwater head at the coastal control dams to prevent saltwater intrusion (Kohout and Hartwell, 1967).

Drainage canals have not only lowered the freshwater level about 5 or 6 feet, but they have also provided a direct channel for seawater to move inland. The problem of salt intrusion has been partially solved by the installation of control structures in most canals. The problem of maintaining sufficient freshwater head in times of drought still remains.

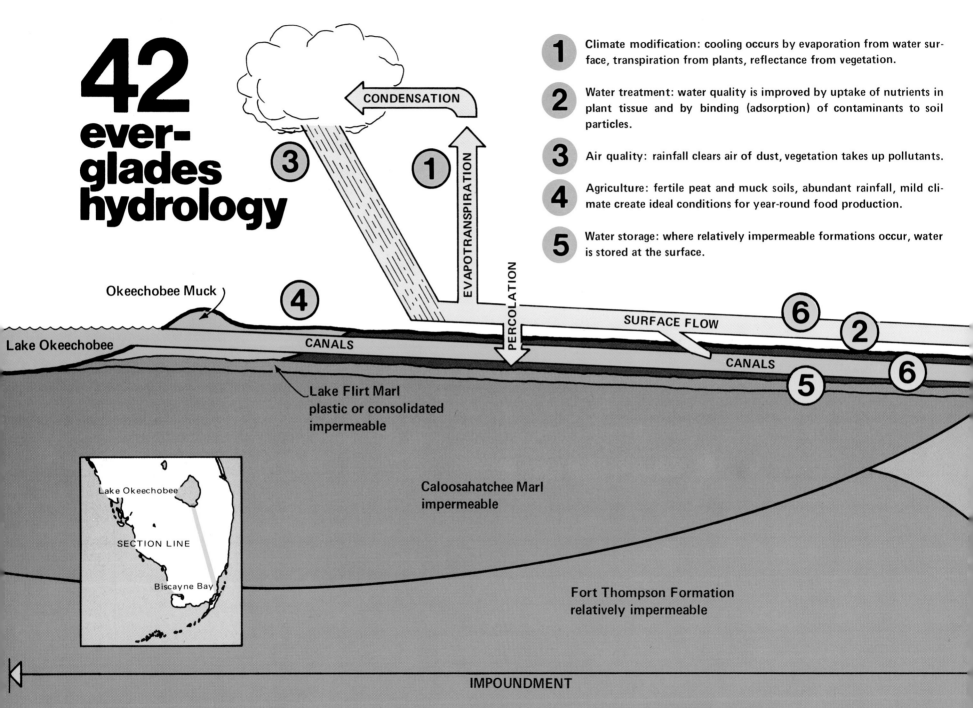

42 ever-glades hydrology

1 Climate modification: cooling occurs by evaporation from water surface, transpiration from plants, reflectance from vegetation.

2 Water treatment: water quality is improved by uptake of nutrients in plant tissue and by binding (adsorption) of contaminants to soil particles.

3 Air quality: rainfall clears air of dust, vegetation takes up pollutants.

4 Agriculture: fertile peat and muck soils, abundant rainfall, mild climate create ideal conditions for year-round food production.

5 Water storage: where relatively impermeable formations occur, water is stored at the surface.

CONDENSATION

EVAPOTRANSPIRATION

PERCOLATION

Okeechobee Muck

Lake Okeechobee

CANALS

SURFACE FLOW

CANALS

Lake Flirt Marl
plastic or consolidated
impermeable

Lake Okeechobee

SECTION LINE

Biscayne Bay

Caloosahatchee Marl
impermeable

Fort Thompson Formation
relatively impermeable

IMPOUNDMENT

SOURCES: Davis, 1943; Parker et al., 1955.

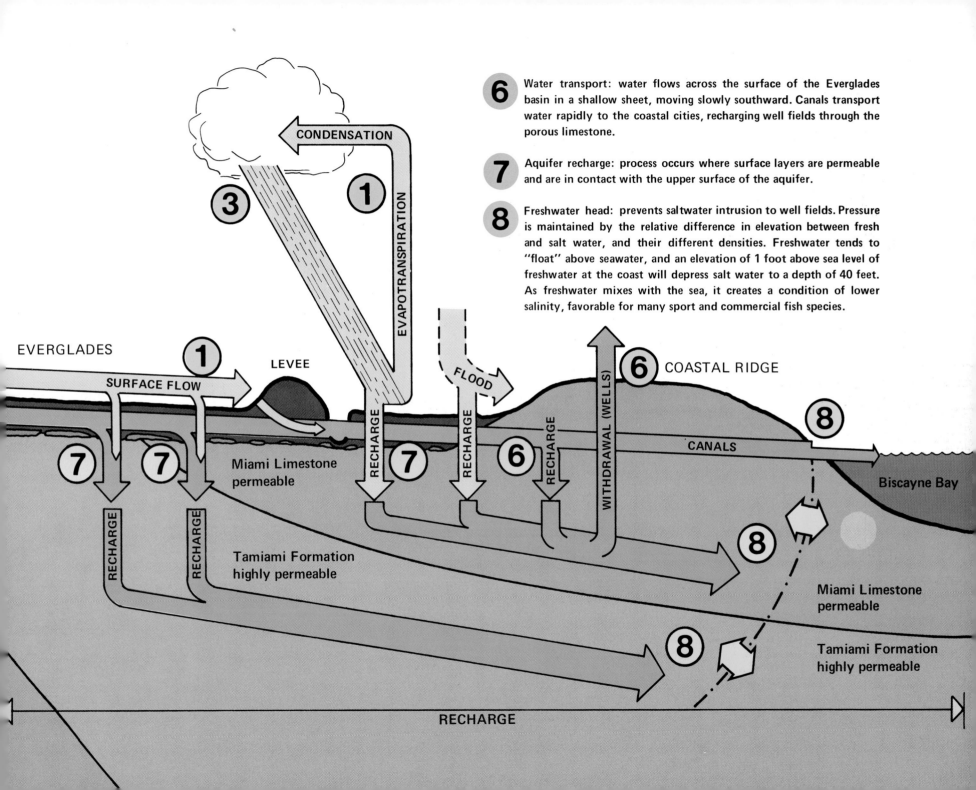

6 Water transport: water flows across the surface of the Everglades basin in a shallow sheet, moving slowly southward. Canals transport water rapidly to the coastal cities, recharging well fields through the porous limestone.

7 Aquifer recharge: process occurs where surface layers are permeable and are in contact with the upper surface of the aquifer.

8 Freshwater head: prevents saltwater intrusion to well fields. Pressure is maintained by the relative difference in elevation between fresh and salt water, and their different densities. Freshwater tends to "float" above seawater, and an elevation of 1 foot above sea level of freshwater at the coast will depress salt water to a depth of 40 feet. As freshwater mixes with the sea, it creates a condition of lower salinity, favorable for many sport and commercial fish species.

CONDENSATION

3

1

EVAPOTRANSPIRATION

EVERGLADES

1

SURFACE FLOW

LEVEE

FLOOD

6 COASTAL RIDGE

WITHDRAWAL (WELLS)

8

RECHARGE

7

7

7

RECHARGE

RECHARGE

RECHARGE

RECHARGE

7

RECHARGE

6

RECHARGE

CANALS

8

Biscayne Bay

Miami Limestone permeable

Tamiami Formation highly permeable

8

Miami Limestone permeable

8

Tamiami Formation highly permeable

RECHARGE

the everglades

The Everglades, "the River of Grass," is a flat expanse of freshwater wetland interrupted by scattered tree islands. Lake Okeechobee lies at the head of this shallow, water-filled basin. Before drainage canals were constructed, the water from the lake overflowed into the northern Everglades, and rainfall augmented its slow overland flow to the south. Below the present Tamiami Trail the flow was channeled into two major natural drainageways. The smaller of these, Taylor slough, still empties into the coastal marsh west of Homestead; the Shark River slough is a larger system that feeds the rivers of the mangrove fringe near Cape Sable.

Around the turn of the century the potential of the rich muck and peat soils was recognized, and a drainage program was begun to permit agricultural development. The Caloosahatchee River on the west and the St. Lucie River on the east were extended to Lake Okeechobee to lower the water level of the lakes. Other canals were dug southward from the lake across the Everglades and through the coastal ridge. These efforts were so effective that salt intrusion occurred at the coast and drought and fire in the Everglades soon followed. Yet, flood waters were still a hazard in wet years, especially when hurricanes brought torrential rains.

In response to these problems the U.S. Army Corps of Engineers began the construction of the three conservation areas. Water is pumped into these large storage areas which are enclosed by dikes and levees. This system intercepts and diverts the surface water flow. Canals connect these reservoirs to the well fields of the coastal cities and to the sea. These flood control measures have considerably reduced the areal extent of the historic Everglades, but much of it still remains as wetland although altered by the change in quantity and seasonal rhythm of water flow. Ponds now stand where a river once slowly flowed.

The northern Everglades basin is now largely devoted to agriculture. The conservation areas occupy the western half of the central and southern portion, with agriculture to the east and urban development spreading from the coastal ridge. The extreme southern part is now Everglades National Park.

Surface Water Storage in the Northern Everglades

The water storage capacity of the Everglades basin is dependent on the characteristics of the underlying limestones. The northern basin is well suited to surface water storage and flow, while the southern portion is an area of recharge for the Biscayne Aquifer.

Lake Okeechobee and the Everglades north of Fort Lauderdale lie over the northern portion of the Fort Thompson formation. In this area the formation is of low permeability, holding the water above its surface rather than allowing it to percolate through. In addition, the surface of the limestone is covered by a thin layer of freshwater marl, a clayey deposit that effectively fills and seals any cavities in the rock. Peat and muck deposits lie over this marl layer. These organic deposits are more permeable vertically than horizontally. They act as a sponge, absorbing large amounts of water at the beginning of the wet season, and only later, when the peat is saturated, does overland sheet flow begin. After the rains cease in the winter the peat remains wet, slowing giving up its moisture to plant roots and to the atmosphere by evaporation. This water storage capacity extends the period when moisture is available for plant growth well into the dry season. The northern Everglades and Lake Okeechobee are ideally suited for water storage at the surface, and their slight slope to the south moves the water slowly in that direction. Canals cut through this region also transport the stored water to the coastal cities.

East of the northern Everglades the Anastasia formation, overlain with Pamlico sand, forms the coastal ridge. This formation is highly permeable, and in many places can yield large quantities of water to wells. Where the Everglades basin abuts the coastal formations, or in the transverse glades, there is a good potential

for groundwater recharge to occur by surface water of the wetlands in addition to that which is supplied by rainfall. This potential water supply is of significant importance to the cities from Fort Lauderdale to Palm Beach, but unfortunately the area is under development pressure. Serious consideration should be given to land use controls that will prevent contamination or saltwater intrusion of this water source.

Groundwater Storage in the Southern Everglades

The southern Everglades are not efficient as a surface water reservoir since the underlying rock is permeable and is not sealed by a marl layer beneath the peat. Instead, this is a recharge area for groundwater storage in the Biscayne Aquifer. This aquifer extends from southern Palm Beach County into Everglades National Park, and from just beyond the western borders of Dade and Broward counties to the east coast. It is wedge-shaped, 100 to 200 feet deep at the Atlantic coast, thinning out to a feather edge on its western and northern boundaries. The aquifer is composed of the Miami limestone and the Fort Thompson formation. Both formations are riddled with solution holes, which makes them highly permeable, one of the most permeable aquifers ever tested by the U.S. Geologic Survey.

Beneath the southern Everglades the eastern third of the aquifer is more perme-

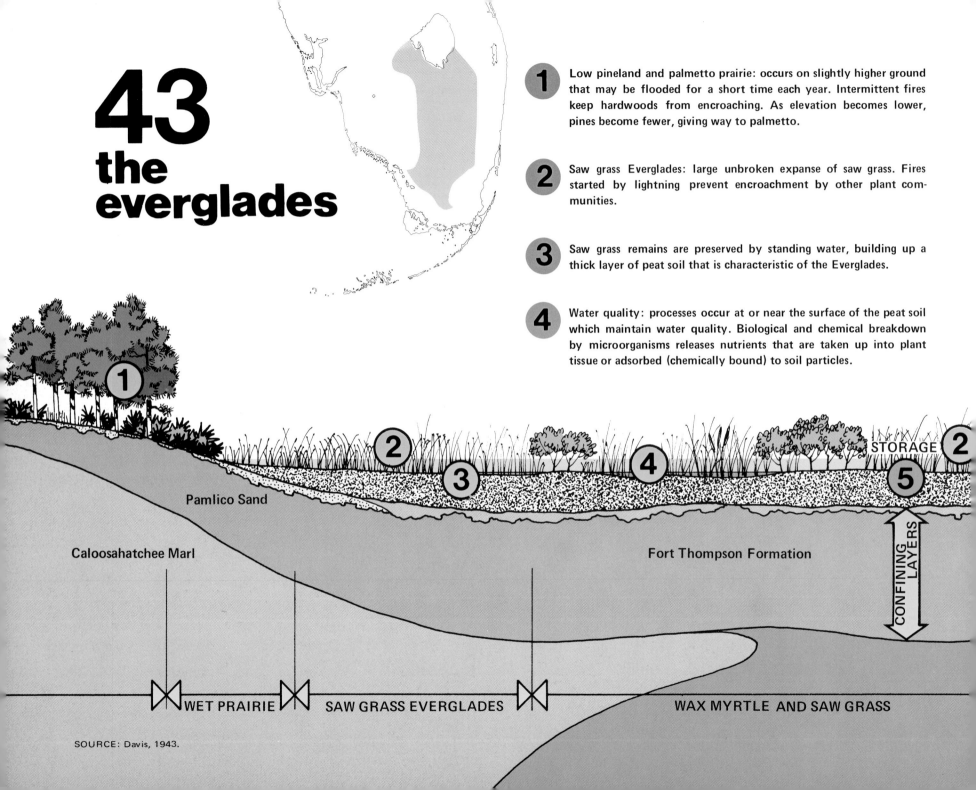

43 the everglades

1 Low pineland and palmetto prairie: occurs on slightly higher ground that may be flooded for a short time each year. Intermittent fires keep hardwoods from encroaching. As elevation becomes lower, pines become fewer, giving way to palmetto.

2 Saw grass Everglades: large unbroken expanse of saw grass. Fires started by lightning prevent encroachment by other plant communities.

3 Saw grass remains are preserved by standing water, building up a thick layer of peat soil that is characteristic of the Everglades.

4 Water quality: processes occur at or near the surface of the peat soil which maintain water quality. Biological and chemical breakdown by microorganisms releases nutrients that are taken up into plant tissue or adsorbed (chemically bound) to soil particles.

STORAGE

Pamlico Sand

Caloosahatchee Marl

Fort Thompson Formation

CONFINING LAYERS

WET PRAIRIE   SAW GRASS EVERGLADES   WAX MYRTLE AND SAW GRASS

SOURCE: Davis, 1943.

5 Water storage: in the northern Everglades, shown here, a layer of marl seals the limestone surface, creating conditions ideal for surface water storage, in addition to storage in the limestone aquifer.

6 Pond vegetation: occurs where water stands all year. Pickerelweed, sagittaria, cattail, water lilies, bladderwort, and other species. The remains of pond vegetation form a mucky peat soil.

7 Willow heads: occur where soil is moist or wet. A pioneer species, willow often invades areas disturbed by recent fire or clearing. Peat builds up beneath the trees, raising the land elevation and providing a habitat suitable for other successional plant communities.

8 Slough: drainage depression which forms a major component of the water transport system of the Everglades. Surface flow moves water slowly southward.

9 Tree island: elongated island of trees in a "sea" of saw grass. They are called willow heads, bayheads, palm glades, or tropical hammocks, depending on the dominant tree species.

10 Aquifer recharge: occurs where pervious formations occur beneath the peat, allowing water to seep into the porous limestone of the aquifer and to be stored there.

11 Low hammock: occurs in the transition zone between wetlands and high hammock or pineland. Water oak, sabal palm, bay, strangler fig are typical species.

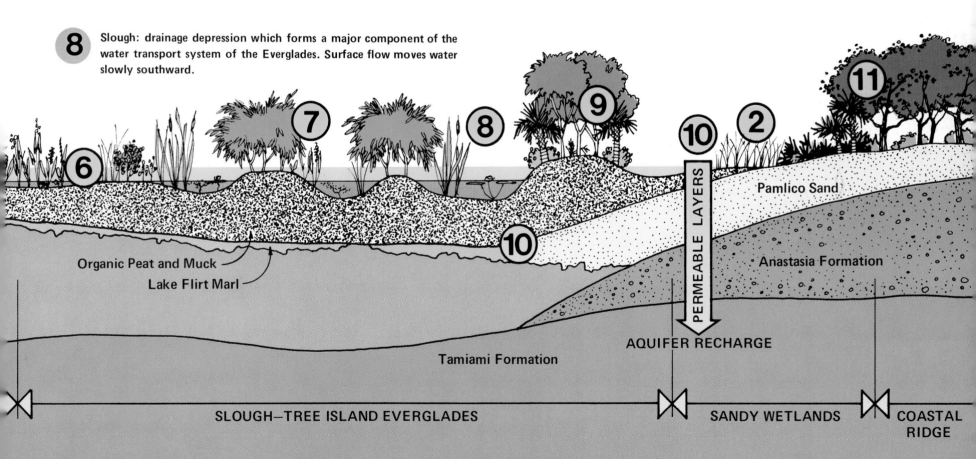

Organic Peat and Muck

Lake Flirt Marl

Pamlico Sand

Anastasia Formation

PERMEABLE LAYERS

AQUIFER RECHARGE

Tamiami Formation

SLOUGH—TREE ISLAND EVERGLADES

SANDY WETLANDS

COASTAL RIDGE

Tree islands in the Everglades

Spider lily

able. To the west it becomes thinner and less permeable, generally because the Fort Thompson formation is denser there and the sandy sections contain a finer sand. The Miami limestone, however, becomes more permeable to the west, and it is through this thin upper layer of rock that groundwater flows to the east to recharge the deeper portions of the aquifer. This permeable layer also allows a large amount of seepage to occur through the aquifer beneath the levees of the conservation areas. The amount varies over time, according to the water levels on either side of the levee (Parker, 1955).

Although the Biscayne Aquifer and the Everglades system provide a huge reservoir of water, the supply is finite. The values of the hunting, fishing, and other outdoor recreation opportunities of the Everglades

are an integral part of the attractiveness of the southern Florida life-style. Tourist dollars are an important part of the economy. Population growth, largely a result of the pleasant outdoor environment, competes for the water that maintains that environment. Better water management techniques are needed to ensure that both a healthy economy and a quality environment can survive.

Vegetation

The Everglades region is a vast expanse of saw grass marsh and wet prairie interspersed with scattered tree islands. Saw grass marshes are the dominant plant community, occurring over areas where water stands all or part of the year. A small difference in elevation can change the plant community completely. Where the water

is slightly deeper the saw grass gives way to emergent pond vegetation, such as spatterdock, pickerelweed, or cattail. When the land surface is slightly higher than the saw grass marsh, tree islands occur. On the higher limestone outcrops subtropical hammocks form, containing hardwood vegetation quite similar to that found in the hammocks of the coastal ridge. Other tree islands grow on mounds of peat. These may be bayheads, containing not only the red bay and white bay, but also dahoon holly, wax myrtle, coco plum, and myrsine. Willow may be the predominant tree in other tree islands, especially in disturbed areas. Ferns are abundant in both hammocks and bayheads.

The tree islands are generally aligned with the direction of water flow that existed in the historic Everglades. This is

Everglades pond

Common egret

Pinnacle rock, an eroded limestone

particularly apparent in aerial photographs. It is not certain if water flow alone is the determining factor, or if some underlying geologic structure defines both water flow and tree island orientation.

Seen from the air, the typical tree island has a blunt, rounded end facing the water flow with an elongated middle portion and a "tail" extending from the downstream end. Often there is a small hammock forest on a limestone outcrop at the blunt end, with the elongated middle and tail being dominated by bayhead vegetation. Many tree islands contain a central pond, and the islands are usually surrounded by a moat that protects from fire and gives refuge for wildlife during the dry season. The moat is created by acidic runoff from decaying vegetation, which dissolves the surrounding limestone.

The pond community of the Everglades is an essential link in the life cycle. During the rainy season fishes and other aquatic animals attain maximum populations. When the rains cease and the waters slowly recede, these aquatic inhabitants become concentrated in the deeper ponds and depressions. Some of these ponds are dug and maintained by resident alligators. Others are solution holes in the limestone. It is in these holes and ponds that enough small aquatic animals survive the dry season to repopulate the Everglades when the rains begin again and water spreads across the land surface. The reproductive cycles of large wading birds, like the wood stork, as well as amphibians, large fish, and mammals are timed to coincide with the period of receding water. As the huge population of small aquatic animals be-

come concentrated in the ponds, the larger animals are able to find and catch food for their voracious young.

Soils

The peat and muck soils of the Everglades are the partially decomposed remains of various aquatic and marsh plants, particularly saw grass. As long as the soils are covered with water, decay is slowed and the vegetative remains continue to accumulate. If the areas are drained, their organic deposits begin to subside due to compaction, oxidation, and bacterial action, at a rate of 1 foot in ten years. Approximately 375,000 acres of peat and muck soil south and southeast of Lake Okeechobee are now being used for pasture, vegetables, and sugarcane. The U.S. Department of Agriculture (1973) estimates that this organic land will essentially disappear by the year 2020.

During periods of drought, peat soils in the Everglades often burn. Ordinarily, fires set by lightning are a vital part of the ecosystem, preventing encroachment of other types of vegetation into the saw grass. When water levels are high, or if the peat soil is moist, only the old tops of the saw grass plants burn and new green shoots quickly sprout from the roots, attracting browsing deer. When the peat is dry, it ignites readily, and it burns until the rains begin again or until the peat is consumed down to rock. There are many such bare rock areas in the Everglades and the Big Cypress.

The spike rush invades areas of burned-out saw grass. Once the water rises above the surface again, it tends to be alkaline and calcium laden because it is in contact with the bare limestone. (Water standing over organic soils tends to be acid). Associated with the spike rush is the blue-green algal mat, an inconspicuous but important component of the ecosystem. It is also a land builder, precipitating calcium carbonate from the water to form a marl deposit. This begins the soil-building process once again, and eventually saw grass will again invade to cover the marl with peat.

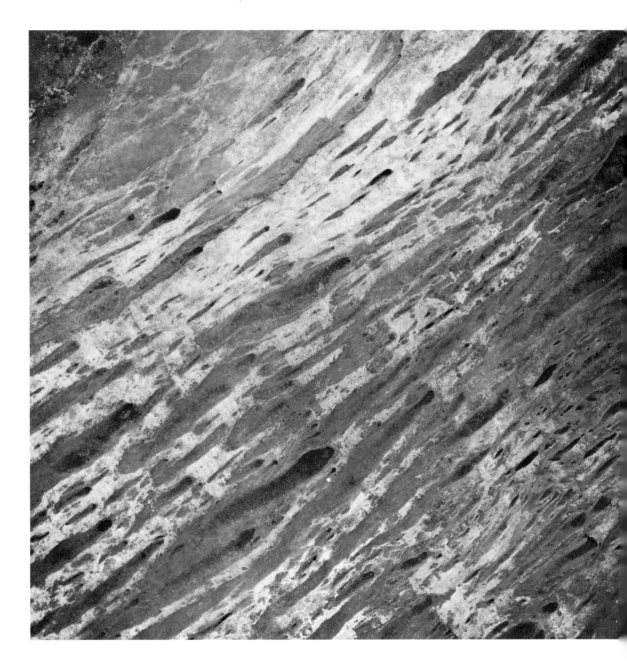

Tree island Everglades. This aerial photograph shows a portion of the Sharp River slough approximately four miles square. A typical tree island is aligned with the direction of water flow, with a blunt rounded end facing upstream and an elongated tail extending downstream.

108

big cypress swamp

The Big Cypress lies to the west of the Everglades on slightly higher terrain; a low limestone ridge divides the two regions into separate watersheds. While the Everglades is a nearly unbroken expanse of wet grassland, the Big Cypress is more complex with grasslands and hammocks interspersed among the cypress forests. It is a water-dominated region, and on the basis of its internal drainage patterns the U.S. Geological Survey has divided the Big Cypress watershed into three subareas (Klein et al., 1970). Subarea A occupies the northeast corner of Collier County, where a low ridge diverts water flow toward the southeast into Conservation Area 3. Subarea B includes 550 acres of the western portion of the watershed and is characterized by man-made drainage associated with the most extensive residential development in the swamp. The central portion of the swamp, subarea C, has been changed very little by human activity, with the exception of the Barron River and Turner River canals. Lumbering activity has had its greatest impact in the Fakahatchee Strand.

Drainage

The western portion of the Big Cypress has been most altered by drainage canals, beginning in the 1920s when the Tamiami canal and the Barron River canal were dug to supply fill for the adjacent roads. In the late 1950s the Turner River canal was dug to supply fill for the roadbed of State Road 840A. These canals are really shallow borrow pits and do not penetrate very deeply into the aquifer. In the 1960s levees and canals were constructed to divert water from subarea A into the Conservation Area 3; the Everglades Parkway (Alligator Alley), which bisects the swamp from east to west, was completed in 1967. As drainage canals lowered the water levels and the roads provided access, large-scale residential land development was begun in the western part of the Big Cypress in Golden Gate Estates. The historic sheet flow and flooding within the development have been largely eliminated, and drainage is directed to the south by the Faka-Union and Henderson Creek canals and to the west by the Golden Gate canal (also see Figure 18).

Most of subarea C, which is still relatively undisturbed, is now included under some form of government control or ownership with the objectives of: (1) protection of the shallow aquifer, which is essential as a source of municipal water for urban development in subarea B and the coastal cities to the west; (2) protection of public investment by ensuring a continued flow of surface water to Everglades National Park; (3) protection of the estuarine fisheries by maintaining seasonality, quantity, and quality of water flow to the Ten Thousand Islands; and (4) mainte-nance of the flora and fauna of this unique ecological region. Most of subarea C west of the Barron River canal is to be acquired by purchase for inclusion in the National Freshwater Preserve (see Figure 23). The state of Florida will probably soon acquire lands in the Fakahatchee Strand, as well as the mangrove wetlands in the Ten Thousand Islands from Marco Island to Everglades National Park. These state and federal lands, and lands in the Okaloacoochee slough and part of subarea A, are designated as an area of critical state concern and are subject to development regulations that are designed to ensure their continued functioning as a water resource.

Geology and Water

The water resources of the Big Cypress watershed are as essential to the future of southwestern Florida as the Everglades and Biscayne Aquifer are to the southeast. An extensive shallow aquifer underlies most of the watershed and extends slightly north into Hendry and Lee counties. It is thickest near the coast where it is composed of Pamlico sand, the Anastasia formation, and Tamiami limestone. Toward the north and east the aquifer becomes thinner, and it tapers out near the Dade and Broward county lines. Areas of low permeability occur in the west, north, and east where there are beds and lenses of fine sand and sandy clay, but the central and southern parts are predominantly permeable, solution-riddled limestone of the Tamiami formation from 60 to 85 feet deep. The upper layers are denser and less permeable than the lower part. Shallow canals do not remove as much water from the aquifer as they would in a more permeable surface formation that would allow more rapid flow through the rock (Klein, 1972).

The yearly cycle of summer rain and winter dry-down duplicates the water regime in the Everglades. Recharge occurs over the entire region by infiltration of rainfall. The water in the aquifer rises as the rainy season continues until the water table reaches the surface and overland flow begins. During the wet season, in areas not drained by canals, 90 percent of the land surface is inundated. Natural drainage is toward the south through sloughs and swamps to the mangrove-bordered estuaries of the Ten Thousand Islands. During the dry season the water recedes until only 10 percent of the land is wet, and water stands only in the deeper sloughs and ponds. Wildlife and plants are adapted to this cycle and depend on it for survival.

Vegetation

Water shapes the elevation-dependent vegetative communities of the Big Cypress. Davis (1943) called the topography "confused"; it is intricate, creating a region of great diversity. Low limestone ridges and shallow sloughs dissect the land, interspersed with numerous ponds and rock outcroppings; the distribution of different types of vegetation follows the topography. Except for the characteristic cypress forests, most of the plant associations have been described in other sections and will only be mentioned here.

Pine islands and hardwood hammocks occupy the higher outcrops. The pines here grow larger than those on the coastal ridge and often form a mixed forest, with cypress growing in the lower potholes of the limestone. The hammocks, best developed in the eastern part of the watershed, contain most of the species found on the ridge. Both pine and cypress woodlands are dotted with grass prairies; bayheads, pop ash, and pond apple populate the depressions. The cypress and swamp forests give the region its name.

Dwarf Cypress. Open forest of small cypress trees probably occupy more land area than any other plant community of the Big Cypress Swamp (Craighead, 1971). The trees grow on marl soil three to six inches deep, with little organic material present. The lack of nutrients stunts their growth, and some trees that are up to 100 years old are no larger than 25-year-old trees that grow in nearby sloughs where soils are deep. There is not enough litter to sustain an intensive fire, and the dwarf cypresses can survive in a dry year when the deeper peats may burn and the forests of larger trees are destroyed.

Cypress has critical requirements for establishment. The seeds sprout in the winter on moist organic soil after the high water recedes, and the seedlings must grow rapidly in the first year to remain above the level of the next summer's flooding. Short periods of inundation can be tolerated, but extended submergence will kill the young tree. On the other hand, prolonged dry conditions, resulting from a lowered water table, can also be fatal.

Cypress Strands and Domes. Cypress heads, or domes, contain the tallest trees in the center, where the peat soil is deepest; near the edge, where soils are shallow, the trees grade down to dwarfs of the same age. The peat surface is seldom higher than surrounding elevations, but merely fills a depression in the bedrock. Often there is a central pond. Strands are similar in structure, except that they occur in elongated drainage depressions. In the shaded interior of the dome, few other species are found there, except for bromeliads, ferns, and orchids. Other marsh species are found near the edges where there is more light. Buttonwood, red bay, coco plum, red maple, and willow are typical (Craighead, 1971). The larger cypress trees were logged in the 1950s, and in many strands the formerly secondary swamp species, primarily red maple and oaks, now dominate (Carter, et al., 1973).

Soil

Soils in the Big Cypress are thinner and sandier than Everglades soils, and there are large areas where the limestone of the Tamiami formation is exposed at the surface. A thin layer of marl or sand is found just under the prairies and dwarf cypress, while peat fills the deeper depressions in the bedrock. The soils become deeper near the mangrove fringe (Craighead, 1971).

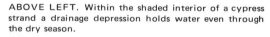
ABOVE LEFT. Within the shaded interior of a cypress strand a drainage depression holds water even through the dry season.

ABOVE RIGHT. Open forests of dwarf cypress probably occupy more land area than any other plant community of the Big Cypress. These trees may be very old, but thin soil and lack of nutrients have stunted their growth.

BELOW. Cypress heads, or domes, contain the tallest trees in the center, where the peat soils are deepest. Near the edge of the dome, soils become thinner and the trees are smaller, giving the forest its typical shape and name.

44 the big cypress swamp

1 Prairies: mostly grasses and sedges, but little saw grass. Occasional sabal palm groves.

2 Dwarf cypress: scattered sparsely in some prairies and saw grass marshes. Dwarfing is caused by thin soil, short hydroperiod, and frequent fire. A tree 3 inches in diameter may be 50 years old.

3 Saw grass marsh: occurs in areas slightly lower than prairies. Saw grass remains form peat, but deposits are not as deep as in the Everglades.

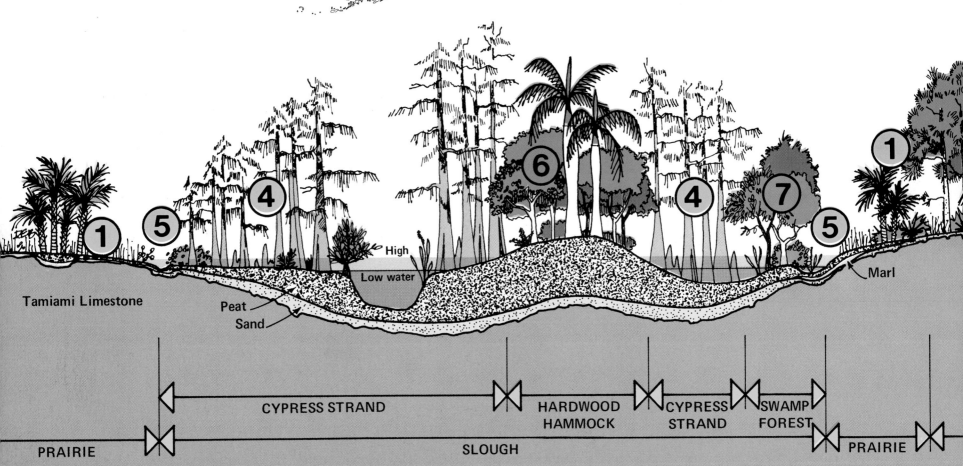

Tamiami Limestone

Peat
Sand

High
Low water

Marl

PRAIRIE — CYPRESS STRAND — SLOUGH — HARDWOOD HAMMOCK — CYPRESS STRAND — SWAMP FOREST — PRAIRIE

SOURCES: Carter et al., 1973; Davis, 1943; George, 1972; Little, Schneider, and Carroll, 1970; U.S. Dept. of Interior, 1969.

4 Cypress dome: swamp forests of large cypress trees occur in drainage depressions where water stands during the wet period. Larger trees occupy the center, grading down to dwarf size at the edge; this creates the domed appearance. Tree branches are often covered thickly with epiphytes, some unique to the Big Cypress.

5 Moat: most tree islands and cypress heads are protected from fire by a surrounding moat etched in the limestone by acidic water.

6 Hardwood hammocks: occur on elevated spots where flooding is minimized. They are protected from fire by the adjoining slough. Gumbo limbo, wild tamarind, oak, red mulberry; in the Fakahatchee, royal palm.

7 Swamp hardwood forest: borders the cypress strand in some places. Magnolia, bay, red maple, coco plum.

8 Pine forest: on higher, rocky ground. Frequent fires keep out the hardwoods. Saw palmetto, sabal palm, and many species of grasses and shrubs.

9 Willows: border ponds and streams, often invading new ground after fire. Peat builds up, provides habitat for successional species.

10 Alligator pond: some heads contain a central pond, bordered by cattails, alligator flag, and other emergent aquatics. The center is kept open by a resident alligator. This is an important refuge and source of food for other animals, fish, and birds in time of drought.

11 Bayheads: Red bay, sweet bay, wax myrtle, holly, coco plum are typical. The bayhead is often surrounded by a ring of sabal palms.

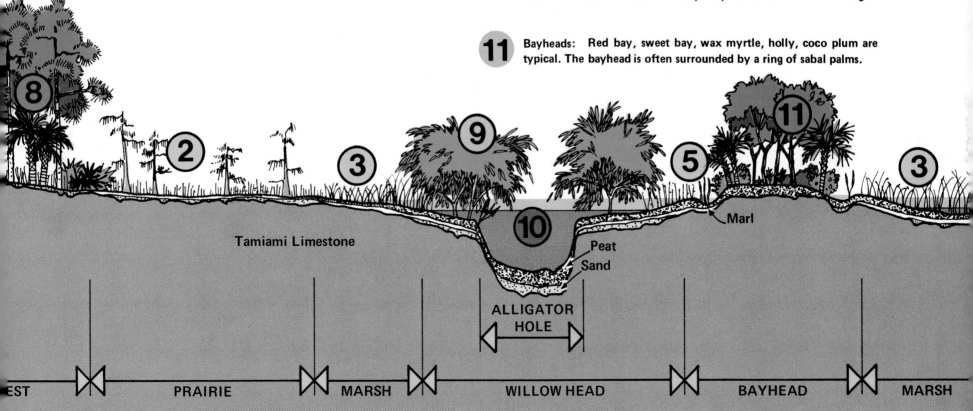

summary

Environmental Services

Water storage and aquifer recharge for supply to urban areas

Storage and dispersal of flood waters

Water quality maintenance by biological and chemical action of vegetation and soil

Climate modification. Evaporation, reflectance, transpiration, and air movement in the wetlands provide an efficient cooling mechanism for the entire region.

Slow overland flow to the coast supplies nutrients and freshwater to coastal ecosystems and fisheries

Surface water and groundwater flows from the wetlands to the coast, preventing saltwater intrusion

Constraints

Drainage of wetlands decreases freshwater storage capacity, both at the surface and in the aquifer. It may also allow saltwater intrusion at the coast.

Alteration of wetlands results in loss of wildlife and recreation values.

Filling affects water quality by destroying natural vegetation, reducing percolation, and increasing surface runoff.

Impermeable surfaces, such as paved areas or buildings, can concentrate polluted runoff and contaminate the aquifer. They also reduce the area available for aquifer recharge.

Improper disposal of solid and liquid waste in wetlands can contaminate the aquifer. Septic tanks are unsuitable as a method of disposal. Central treatment facilities should be located above the 100-year flood level.

Development in the wetlands is subject to severe flood hazard.

Deep wells into the brackish Floridan Aquifer can contaminate the shallow freshwater aquifer and surface waters if the wells are not properly cased and sealed.

A depleted aquifer may ultimately be the limiting factor that will curtail urban growth in southern Florida. Therefore, wise use of this resource, maintenance of its quality, and assurance of its renewal, is of prime importance to the continued economic well-being and quality of life of the entire region.

Opportunities

The criteria for delineating an area as suitable for aquifer recharge are as follows:

The geologic formation must be permeable.

Water must be available, either in the form of adequate rainfall or as overland flow. Ideally, both would occur.

Maximum possible permeable surface must be preserved. This implies low surface coverage by impermeable materials, such as paving or buildings.

The areal extent of the recharge zone must be sufficient to balance depletion of water resources from evaporation from water bodies, runoff from canals, and urban use.

The area should be in relatively pristine condition. Least desirable are agricultural lands with their load of pesticides, fertilizers, and animal waste. Areas near sewage outfalls are obviously unsuitable, and most canals are polluted both by agricultural and urban runoff, as well as by sewage effluent. Rockpits are not an undesirable element, as long as they are not connected to the canal system (Hartwell, 1975).

The limiting factor to downstream community growth may be the contamination of its water supply by pollution from upstream development.

Designation of an area as an aquifer recharge area does not preclude development. However, careful assessment of local and regional needs for water in the future must be carefully assessed. The areal extent of necessary recharge must be evaluated to match these future needs. Then the recharge areas can be developed as long as certain criteria are met and important constraints are observed:

Sufficient organic soil cover must be preserved to maintain water quality. The filtering action of soil particles and the biological and chemical processes at and beneath the soil surface inactivate many viruses and bacteria, including those which cause typhoid, dengue fever, and the like.

Before development occurs, natural drainage patterns of the site and the surrounding region should be identified and incorporated into the site design in order to maintain overland flow through the development. This objective can be achieved by design solutions, such as stilted buildings or green swale drainage systems that are imitative of natural systems. Where the natural pattern must be interrupted, culverts, bridges, and similar measures should be designed to approximate the natural water regime.

Landscaping should be designed to permit natural growth and attrition, rather than mowing and trimming. Low maintenance plant materials should be used to avoid the necessity for application of fertilizers and pesticides. Native vegetation is admirably suited, since it is well adapted to the South Florida environment and does not require chemical and human energy inputs for survival. Nutrient uptake by vegetation is a significant service performed by the natural environment in the maintenance of wa-

ter quality. If such materials as grass clippings are allowed to return to the soil, however, these nutrients are released back to the water supply.

Use of exotic (nonnative) plant species in landscaping should be avoided. High maintenance of some exotics requires chemical inputs that cause water pollution. Low maintenance exotics, on the other hand, tend to escape from cultivation. They can invade and replace natural vegetative communities, decreasing the system capacity for providing environmental services. Wetlands, especially in disturbed areas, are particularly susceptible to invasion. Maintaining water quality depends on a diverse and complex system of plants, animals, and microorganisms. When exotics are introduced, a simple system results that lacks the effectiveness and buffering provided by a more diverse system.

mangrove and coastal marsh

On low-lying tropical shorelines around the world mangrove forests fringe the sea. In southern Florida they occur intermittently along the entire coast, primarily in quieter estuarine waters. Sometimes, as on the Keys, mangroves face the high energy of the open sea, as long as there is a protective near-shore area of shallow water. One of the largest and best developed mangrove forests in the world occurs in the Ten Thousand Islands along the southwestern Florida coast (Davis, 1943). Coastal marshes are found adjacent to to the mangroves, but are more common in the northern part of the Florida peninsula where frost limits the spread of mangroves. In southern Florida, both mangroves and coastal marshes cover large areas, sharing the important characteristic of being able to withstand either intermit-tent or constant contact with seawater—a rigorous environmental stress that kills most terrestrial vegetation.

Zonation

The term "mangrove" refers to tree species not closely related botanically but generally found in close association in the mangrove forest. Although there are many species of mangrove, only three are found in southern Florida. These are the red mangrove, **Rhizophora mangle;** the black mangrove, **Avicennia nitida;** and the white mangrove, **Laguncularia racemosa.** Classically, they occupy distinct zones within the forest, depending on the degree of salinity and length of inundation that each species can tolerate.

Red mangroves occupy the outer or seaward zone. They are distinctive in appearance, with arching prop roots that project from the trunk or branches down into the water. They produce cigar-shaped seedlings that sprout while still attached to the parent tree, giving the new growth a head start—a competitive advantage in the harsh saline environment. The seedlings are sharply pointed and weighted at the bottom end. After they drop into the water and float away they can easily become rooted if they are caught in debris or in a mudbank in shallow water. Red mangroves can grow in shallow offshore areas where the roots are always under water or onshore where the soil surface is barely under water at the highest tide.

The seedlings can also be carried far inland, into the coastal marsh, where they may grow in dwarfed form in isolated clumps, or up tidal rivers where they form tall forests in the floodplain (Lugo and Snedaker, 1974). These environments vary considerably in salinity, both with respect to tidal flushing and seasonal rainfall. Although mangroves tolerate salt water, it is not necessary to their survival.

The middle zone, at slightly higher elevations, is dominated by the black mangrove in association with salt marsh plants. This zone is usually inundated at high tide, but is otherwise exposed. The roots produce pneumatophores (fingerlike extensions above the soil surface) described by Davis (1943) as resembling a patch of asparagus tips. Black mangroves may be found in pure stands in shallow basins where seawater remains standing between tides. The heat from the sun evaporates some of the water, leaving highly concentrated salt water behind. In the Keys during the dry season, salt becomes so concentrated that crystals sometimes form on the soil surface. These observations would indicate that black mangroves are the species most tolerant to high soil salinity, although even they can be stunted or killed if salt content becomes too great (Teas, 1974). Black mangroves may sometimes be found at the edge of the sea, with no intervening band of red mangrove. Often these trees are old and large, remaining after the reds have

been ripped away by some storm that eroded the coastline.

The most landward zone is affected only by the highest spring and storm tides. The white mangrove is found here in association with the buttonwood (not a mangrove), as well as with transitional marsh vegetation. In some places this zone defines an embankment called the Buttonwood Levee by Craighead (1971). This levee may be a remnant of an ancient shoreline; it impounds and slowly releases the surface flow of fresh or brackish water from the upland marsh into the mangrove system.

Behind the innermost mangrove zone, in areas subjected at times to tidal flooding, the salt marsh forms a transition zone, merging with freshwater marsh vegetation as the salinity of the soil decreases toward the upland. This is an area of change. In a series of years with high rainfall the freshwater vegetation may advance seaward, but during a dry cycle the salt marsh moves back toward the upland. Rising sea levels can make the change a permanent one. On the seaward side mangroves may invade; but, if a hurricane destroys the mangroves, salt marsh vegetation will soon cover the former forest floor. Plants of the salt marsh include saltwort, glasswort, and sea daisy near the mangrove fringe and grade into grasses and other low herbaceous growth that is almost indistinguishable from similar freshwater species, except to the trained

observer (Davis, 1943). This area of subtle change beyond the salt marsh is called the coastal marsh. It does not have a sharp line of demarcation between salt and line of demarcation between salt and freshwater, but rather, it is a transition zone. Usually the water is fresh or brackish, but the vegetation is adapted to sporadic flooding by hurricane-driven ocean waters.

Mangrove Forest

The mangrove system is a highly productive food source for economically valuable commercial and sport fish. It is seven times as productive as an alfalfa field and twice as productive as a corn field, according to biologist W. E. Odum (1971). Just as livestock production depends on alfalfa and corn, so fisheries production depends on mangroves and estuarine marshes. The crop, however, needs no human inputs of fertilizer, money, energy, or labor.

Mangroves are only one of several highly productive estuarine systems. Salt marshes, mud flats, and sea grass beds perform similar functions in other food chains. However, since mangroves are so prevalent on the coasts of southern Florida, they will serve as an excellent example of the general value of such systems.

Detritus. Plant litter, such as leaves, stems, and twigs, are disintegrated into small particles by mechanical action or biological breakdown (digestion). These particles, called detritus, are the basis of

Red mangroves

Black mangroves

White mangroves

the detrital food chain, a significant part of many ecological systems.

In his study of a South Florida estuary, Heald (1971) found that 85 percent of the plant detritus in the basin originated from 2,600 acres of red mangroves, with a total estimated production of 12,400 metric tons (dry weight) per year. The remaining percentage came from 400 acres of saw grass and juncus marshes. Red mangrove leaves alone contribute over three tons per acre annually (W. E. Odum, 1971). Once the leaves enter the water, bacteria, fungi, and other microorganisms,

as well as crabs and amphipods, begin to break down the plant tissue into smaller and smaller bits until it is eventually converted to fine particles suspended in the water. These particles account for 80 to 90 percent of the nutrition of detritus feeders—small fishes or various invertebrates, such as worms, shrimp, small crabs, or insect larvae. These in turn are eaten by larger fishes and invertebrates, as well as by wading birds. Approximately one-half of the total annual production of debris, in the form of fine particles, was not consumed in the estuary studied, but

was carried out by currents to adjacent waters to contribute to food chains there.

Effects on Fish Production. Over sixty species of juvenile fishes depend on the mangrove-bordered estuary for some part of their life cycle. Some of these are important commercial or game species. Tarpon, snook, and lady fish enter the estuary as postlarvae. Gray snapper, sheephead, spotted sea trout, and red drum spend several years of their lives in the mangroves, and crevalle jack, the gaff-topsail catfish, and the jewfish are also found there (W. E. Odum, 1971). Shrimp feed

Buttonwood

Coastal marsh with invading red mangroves

directly on detritus, and so may mullet and blue crabs.

The value of commercial landings of species dependent on mangrove estuaries reflects the contribution of mangrove productivity. Sport fishing is an important component of the tourist economy. Statewide, in 1973, commercial yields were:

	Pounds	Value
Mullet	30,085,634	$3,319,425
Sea Trout	3,324,368	1,166,362
Blue Crab	13,511,913	1,678,901
Shrimp	29,197,597	26,247,871

(National Marine Fisheries Service, 1975)

Mangrove Forest Types. The classic pattern of zonation previously described and shown in the cross-sectional diagram is typical in mangrove forests on relatively steep, uniformly sloping shorelines. In southern Florida the zonation is not clearly defined everywhere because of low topography and the many islands, rivers, and embayments within the forest, and also because of disturbances from human activity (Teas, 1974). The principles are the same, however, and the species are still distributed in response to tidal influence. In periods of gradual lowering of the

sea, mangrove zones may move seaward; but, when tidal action is high, or the sea is rising, the mangroves migrate inland.

Snedaker and Pool (1973) have found five mangrove forest types in southern Florida, with distinctive differences in structure. As in the zoned forest, the pattern is strongly related to the action of water, both the frequency and the amount of tidal flushing and freshwater runoff from the upland.

FRINGE FOREST is located on sloping shorelines where elevations are higher than high tide and where tidal water does

119

45
mangrove and coastal marsh

1. **Land building:** marine grasses and mangrove roots slow currents, acting as sediment traps and building up the surface level.

2. **Turtle grass beds:** take up nutrients from upland runoff, support population of small animal "grazers" on their leaves.

3. **Fish nursery:** juvenile stages of commercial and sport fish feed on "grazers" and find shelter from predators.

4. **Storm protection:** wave and storm forces are absorbed and lessened as they pass through roots and foliage. Erosion protection for shoreline is provided by root system.

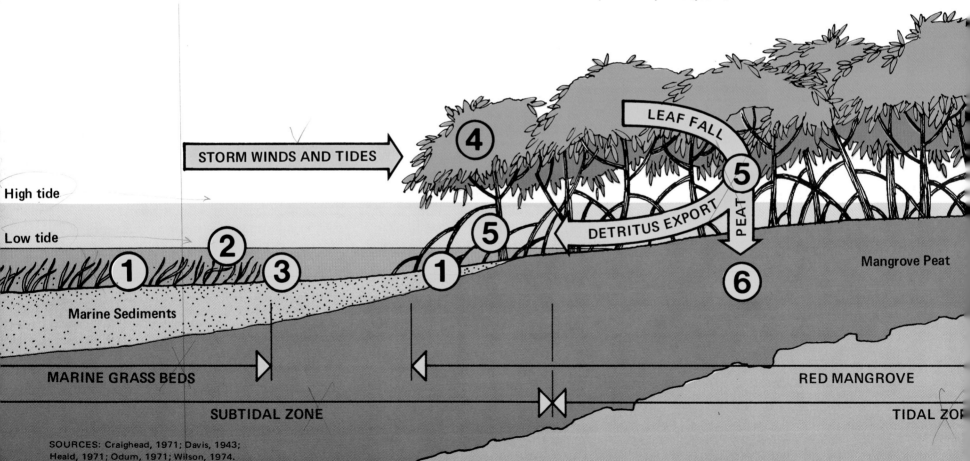

STORM WINDS AND TIDES

LEAF FALL

DETRITUS EXPORT

PEAT

High tide

Low tide

Mangrove Peat

Marine Sediments

MARINE GRASS BEDS

RED MANGROVE

SUBTIDAL ZONE

TIDAL ZO

SOURCES: Craighead, 1971; Davis, 1943;
Heald, 1971; Odum, 1971; Wilson, 1974.

5 Detritus cycle: oysters and crustaceans live on mangrove roots, utilizing mangrove debris and other nutrient particulates. Commercial and game fish find shelter and feed on smaller organisms. Red mangrove leaves fall, decay, provide food for many bacteria, fungi, and small animals.

6 Land building: as sea rises mangrove roots and debris build up layer of peat, in some places 12 feet deep.

7 Black mangrove: root structure stabilizes the soil. This is the most salt-tolerant mangrove species.

8 Buttonwood embankment: impounds upland runoff. Water infiltrates, flows through soil. In the process, water quality is improved.

9 Algal mat: precipitates calcium carbonate from the water to form marl soil.

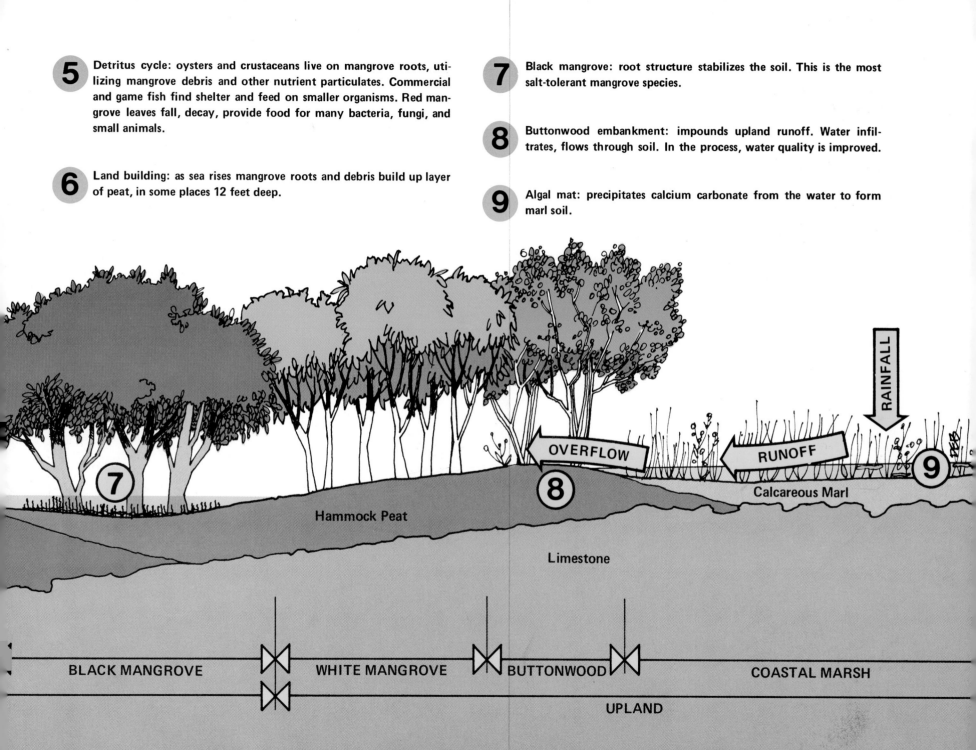

RAINFALL

OVERFLOW

RUNOFF

⑦

⑧

⑨

Calcareous Marl

Hammock Peat

Limestone

BLACK MANGROVE

WHITE MANGROVE

BUTTONWOOD

COASTAL MARSH

UPLAND

not pass completely through the forest but simply moves in and out. The water velocity is not great enough to carry large debris out into the bay, but small particles are flushed out. This forest type most closely resembles the classic zoned forest. Because of its location on exposed shorelines, the fringe forest is vulnerable to damage by strong winds and storm tides.

RIVERINE FOREST occurs along tidal rivers that empty into bays, as in the Ten Thousand Islands. The trees grow on the flood plain, often separated from the river by a berm. Sometimes a fringe forest develops along the slope of the riverbank, with riverine forest behind it. These forests are flushed by tidal rise and fall, but the flow velocity is so low that no litter on the forest floor is redistributed. Nutrient-laden freshwater runoff from the upland during the summer wet season raises water levels and lowers salinity. The riverine forest is dominated by tall, red mangroves with straight trunks and few prop roots. Black and white mangroves are scattered among the reds.

BASIN FOREST also occurs on the mainland but in shallow drainage depressions. Near the coast, under the influence of the daily tides, red mangroves predominate. Further inland, where there is a greater percentage of freshwater in the drainageway, black and white mangroves become plentiful, and all three species support a number of epiphytes, such as orchids and air plants. Where the water becomes completely fresh, the basin forest grades into Everglades vegetation, which can compete better than mangroves in a freshwater environment.

OVERWASH FOREST is dominated by red mangrove on small, low islands and narrow projections of land in shallow bays. They are in the path of tidal flow patterns and are overwashed at each high tide. The incoming tide carries away all loose debris and deposits it in the inner bay.

DWARF FOREST occurs on the flats and in the Florida Keys. The scattered, stunted trees look like juveniles, but they are actually forty or fifty years old. The dwarfing is thought to be the result of extremely low nutrient levels in the soil and water.

Land Building and Erosion Protection. All these forest types are associated with deposits of peat. The leaves that fall on the soil are reduced to detritus in about a year or are washed away by tides. The roots, however, resist decay in the saturated soil in the absence of air, which is necessary for decomposition. These fibrous root remains form a peat that can persist for geological time periods and has been found to depths of 12 feet. These deposits have kept pace with the rising sea at a rate of 3 inches each hundred years (Craighead, 1971).

The complex root structure of living mangroves also acts as a sediment trap. As sediment-bearing water enters the root area, it is slowed down, and the sediments drop out on the bottom. Mangrove peats often have layers of sand or shell imbedded in them, probably deposited by storm tides. These deposits add to the height of the land. The danger exists, however, that too thick a deposit may smother and kill large forest areas. This happened during hurricanes in 1935 and 1960, when the mangrove fringe was pushed back 50 to 100 feet. Large, dead, black mangroves now mark the shoreline, which is receding with wave action and the rising sea. Natural reforestation may be prevented where thick layers of sea grass and seaweed are deposited on the beach by the tides. This excessively thick mat prevents red mangrove seedlings from anchoring and growing (Craighead, 1971).

The mangroves, in their role as buffers to storm winds and tides, prevent devastation of the coastline (Lugo et al., 1971). The energy of storm forces can be absorbed to an extent by friction in densely vegetated areas. This buffering protects the upland, but in the process the mangroves may be damaged or destroyed. In some cases natural recovery takes place, but in others a new shoreline replaces the old. Where upland development is to be protected, some form of management may be necessary.

Mangrove Forest Management. The many valuable functions of mangroves are being revealed by ongoing research, and what was once thought to be a wasteland is being recognized as a valuable component of the regional system. Where possible, mangroves and their associated upland should be left intact to perform these functions in the dynamic cycles of storm and tide, drought and flood. Generation and regeneration occur on wilderness shorelines without the interference of man. Where shoreline systems are impacted by human activity, serious problems arise. Florida mangroves, however, are well suited for careful management practices. The red mangrove is recognized as the most important species for fisheries productivity. It also serves as a soil producer and stabilizer as well as a storm buffer. The most important forest type in these respects is the coastal fringe forest. The riverine and basin forests also provide inputs into detritus production. The black mangrove is important for shoreline stabilization, both as a secondary defense behind the reds, and because its root system

ABOVE LEFT. Large, dying, black mangroves mark a former shoreline which is receding with the rising sea.
ABOVE RIGHT. Black mangrove roots stabilize the soil, trap sediments, provide habitat for small crustaceans.
BELOW LEFT. Red mangrove seedlings sprout roots while still attached to the tree, an advantage for survival in the saline environment.
BELOW RIGHT. Red mangroves in the coastal fringe are vulnerable to damage by strong winds and storm tides.

develops more rapidly than that of the red.

One form of management is the planting of mangroves in new areas. Mangroves were planted to stabilize portions of the causeway of the Florida Overseas Railway. In Asia it is reported that mangroves are closely planted just offshore to trap sediments and thus claim land from the sea. Teas and others have started new plantings in Vietnam, Everglades National Park, St. Lucie County, Charlotte County, and several locations in Dade County. A spoil island in Collier County was planted by other researchers (Teas, 1974).

Experimental work by Savage in Tampa Bay showed that the black mangrove is at least as important as the red in shoreline protection and land building. The black mangrove has the added advantage of being more tolerant to lower temperature as well as to adverse soil conditions, such as may be found in man-made shorelines. It can also tolerate some inundation by pumped materials by producing new sprouts, and it can similarly survive heavy top damage by hurricanes or frost. It is also quicker to develop an accessory root system to protect the soil from erosion. After five years, however, the prop roots of the red mangrove become increasingly functional in substrate protection.

Both species, as well as white mangroves, respond well to pruning, which suggests that they could be used as waterfront hedges, even to protect existing seawalls or possibly replace them (Savage, 1972). Existing artificial shorelines have frequently suffered damage from erosion. Protective structures have proved to be more expensive than enduring or beautiful. Mangrove plantings would be productive, protective, and economic, if incorporated into waterfront development plans or causeway construction.

It may be possible to encourage the growth of the red mangrove by management practices. In Malaysia, selective ditching or channelization is practiced under government regulation to increase the growth and density of red mangroves so they can be used for charcoal or firewood. Teas (1974) also observed that in Biscayne Bay and at Marco Island red mangroves grow taller and closer together along tidal creeks than in nearby areas. Channelization, however, short-circuits the slow overland flow of nutrient-laden runoff from the upland through the mangrove forest to the sea. Using field measurements and a computer model, Lugo, Sell, and Snedaker determined the probable results of reduced upland runoff. When normal runoff was reduced by 50 percent, mangrove production was reduced proportionately. Below 50 percent, production levels were similar to those of scrub mangrove (Lugo and Snedaker, 1974). Possibly, increased growth along artificial channels would be traded off for decreased productivity elsewhere.

Scrub mangroves appear to be limited

in growth by lack of nutrients and poor soil conditions. Experimental work is needed to develop management techniques for increasing growth and productivity. One such possibility is the application of treated sewage effluent to the mangrove fringe. If the effluent were properly dispersed, mangrove productivity would probably increase, with an accompanying increase in the value of commercial and sport fish landings. Effluent is now being applied to golf courses, with good results. It should be emphasized, however, that long-term field experiments are absolutely necessary to ensure that no undesirable changes would take place, either in the mangrove system or in the receiving waters.

Mangroves are much maligned as mosquito breeding areas. Mosquitos do not breed in the intertidal red mangrove zone, which is the most productive part of the forest, but in upland areas where the eggs are laid on damp soil (Robas, 1970).

Coastal Marsh

Although the mangroves reach their best development on the southwest coast in the Ten Thousand Islands, the coastal marsh dominates the southeast in the area between the ridge and the narrower mangrove fringe of Florida Bay, Card Sound, and south Biscayne Bay. Biologist Susan Wilson describes the region as "flat and low, lying less than a foot above mean sea level. Tree islands rise about an additional one-half meter above the lowlands, which in an average season, flood during the summer rainy season and gradually dry down completely in the dry winter season.

"Although the waters are fresh, except in periods of occasional hurricane floods or drought, red mangroves dot the lowlands and fringe the tree island hammocks which harbor bay, buttonwood, mahogany, poisonwood, seagrape, spanish stopper, and saw grass, among others.

"It is in the shallow waters of the lowlands that the mat community thrives. Here spike rush forms a sparse vegetative cover allowing abundant light to penetrate the clear waters to the algal mat which carpets the marl and enshrouds the submerged portions of the rush stems.

"In May or June, following the winter dry period, the rains and the water overflow from Canal-111 gradually reflood the area. As the water returns, the dried mat swells and resumes growth. Two varieties of fish appear—mosquito fish, and the sheephead minnow. The crayfish leave their burrows in the wet marl and begin to feed.

"This 'spontaneous generation' of mosquito fish and minnow may be occasioned by the work of the crayfish. The fishes survive drought in crayfish burrows, which sometimes descend as passageways to the underground water table.

"During the summer months the mat continues to grow, reaching a thickness of about 3.5 cm. In a typical year, as the dry season progresses, the mat dries, shrinks, cracks, and curls along its desiccated edges, forming irregular, leaden colored platters" (Wilson, 1974).

The algal mat is a major producer in this simple ecologic community. It is also a land builder, precipitating calcium carbonate from the calcium-laden waters to form a marl deposit over the limestone. The rate of depositions at present is less than half an inch per year, although it was greater in the past. This deposit has been called Lake Flirt marl, and near the ridge it has been used extensively for vegetable farming, especially for potatoes. The fine, clayey particles of marl seal the porous limestone substrate, preventing rapid infiltration of surface waters. Instead, rainfall and upland runoff slowly flow to the mangrove fringe.

"Although the waters which flow over the mat presently are unenriched, there is ample reason to predict that nutrient and pollutant loadings will increase. Developmental pressures on the lands which drain into the upper reaches of Canal-111 are mounting. It is reasonable to presume that the mat would be capable of trapping nutrients and pollutants. If further experimentation establishes this characteristic, the mat could be shown to be very important in modifying the overland sheet flow of fresh water which eventually reaches Everglades National Park and the rich marine nursery grounds of the red mangrove fringe" (Wilson, 1974).

summary

Environmental Services

Storage and dispersal of flood waters

Filtration of runoff improves near-shore water quality

Protection of upland by buffering water surge and wave energy

Stabilization of shoreline

Land building by trapping of sediments and other materials and by accumulation of mangrove roots; marl deposition occurs in coastal marsh

Habitat and food for many marine species, including sport and commercial fish

Wildlife habitat, especially for waterfowl and wading birds

Constraints

Hurricane and storm flooding potential is high. Federal Flood Insurance regulations require first-floor elevations to be above the 100-year storm level.

Development below mean high water is under state jurisdiction.

Recent state actions have generally disapproved permit applications for development that would destroy significant areas of mangrove or coastal marsh.

Soils are marls or mangrove peats, of low-bearing capacity.

Indiscriminate removal of vegetation and soils will degrade water quality by reducing capacity of the natural system to act as a trap for sediments and nutrients.

Improper disposal of solid and liquid waste degrades water quality. Septic tanks are unsuitable as a disposal method. Central sewage facilities should be located above the 100-year flood level.

Opportunities

A combination of cluster development, elevated construction, and selective clearing affords a development strategy that is both environmentally sound and economically feasible. It preserves the coastal protection function of the natural system and assures that water quality will not be degraded.

Planned unit development should be encouraged to allow necessary flexibility in urban design. Site planning and building location should be responsive to site conditions. Bonus densities for site preservation and setback from the edge of the water could provide incentives.

Development planning should include selective clearing and clustering of building units to maximize preservation of natural vegetation and ground cover. Cluster development with common marina facilities are preferable to conventional canal developments.

Elevated structures for coastal development solve several problems. Soil conditions in mangrove and coastal marsh

FIGURE 46: Design concept for clustering elevated structures.

areas generally require pilings, and flood criteria require the first floor to be elevated above the 100-year flood level, which in some places is 14 feet above mean sea level. These requirements, taken with environmental considerations, suggest that structures elevated on pilings, without filling beneath, would be the best design solution for these conditions.

The economic advantages of elevated construction make it feasible to build. The alternatives of floating foundations, foundation walls, and/or filling to flood criteria elevations add unnecessary cost.

Consideration should be given to light construction and multilevel design to take maximum advantage of the foundation system.

Low-rise buildings should be located nearest the shoreline, with higher buildings located behind them on the upland. This design strategy affords many more units a view of the water and reduces the impact of development on natural coastal systems.

The heavy equipment necessary for construction on pilings can cause a great deal of damage to natural vegetation on the site. Access for equipment should be limited to future road alignments.

The value of mangroves as protection for the coast is considerable. The use of elevated structures, combined with mangrove forest preservation and management, make costly and environmentally damaging bulkheading unnecessary.

Marina facilities should be designed so that tidal flushing will occur and so that the water will pass over a mangrove area before it reaches near-shore waters.

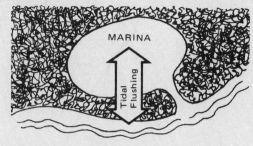

FIGURE 47: Marina.

Selective clearing in mangrove and marsh areas should follow these minimum guidelines to maintain water quality:

Basin forest should be left intact, with an adjacent buffer strip.

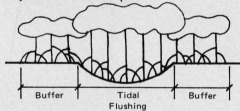

FIGURE 48: Basin forest.

Riverine forest. Where there is some tidal flushing, the tall, red mangroves should be left intact, with a buffer as described for the basin forest. Channels may be cut through to increase tidal flushing. On the other hand, where there is no flushing, the tall red mangroves may be removed without affecting most marine life.

Fringe forest. Along the coast, and bordering canals and streams, fringing mangroves should be left intact in a buffer zone 50 feet inland of the high-water line. These trees may be trimmed to allow a view of the water from upland development. Mangroves seaward of the mean high-water line are protected by state law.

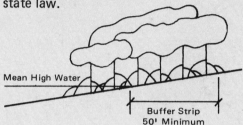

FIGURE 49: Fringe forest.

Dwarf forest. Where growth is very dense, areas bordering the coast and canals should be left intact, as in the fringe forest just described. Tidal channels may be cut through the forest and may actually increase productivity. Where growth is sparse and trees are so far apart that their branches do not touch, dwarf mangroves are not very productive and do not contribute significantly to the environment.

Coastal marsh. Buffer zones similar to those recommended for the mangroves should be preserved in the coastal marsh. A buffer of marshland should be left intact adjacent to water bodies and streams.

127

part three

3

the built environment

the built environ- ment

The purpose of this book is to encourage the use of comprehensive design and planning principles in the creation of the built environment while preserving the integrity and utilizing the opportunities of the natural environment. One's **environment** is defined as the aggregate of all the social, economic, ecological, and physical conditions that influence his life. **Comprehensive** is defined as the consideration of all the factors related to a particular action. **Quality** is simply the satisfaction of all the requirements. It is only through a coordinated and conscientious effort of comprehensive decision making at all levels of public and private involvement that optimum environmental quality can be achieved. The interrelated built:natural environment is a unified web of actions and reactions, held tightly together in a delicate balance of tolerances—social, ecological, economic, political. It is this balance of tolerances that limits the theoretical capacity for any part of the environment to operate beyond that for which it was designed, be it traffic on a road or water levels in the Everglades.

Whether land is considered a commodity in the marketplace or the format for resource conservation, evaluation of the factors affecting each segment of a proposal must be considered throughout the planning process, its implementation, and management. One is directed in day-to-day actions by preselected goals and objectives, and their clear definition is important to minimize diversion from the course. More important, however, is that the design process be applied completely as one proceeds; such adherence will help to ensure the quality of the end product. The process is complicated by overlapping and sometimes conflicting objectives of government agencies and by public and private interests with single-purpose objectives. This section examines the comprehensive planning process as a tool to stimulate meaningful results by those involved in the day-to-day decision making and by those defining long-range policies and goals. Government and private industry have joint responsibility for achieving the highest level of environmental quality for existing and future generations of South Floridians.

"No matter how well we come to

understand the ecosystems upon which we depend, and our relationship to nature, **this understanding must eventually be transferred into planning, management and administrative policies to be of value to both man and nature.** Whereas the leadership responsibilities are well defined, any major revision of policy must be based on a public commitment" (Lugo et al., 1971, p. 97, emphasis added).

Obviously, an ideal development is well adapted to its surroundings, and its position in the spectrum of the built:natural environment not only satisfies but is shaped by its environs and not vice versa. It is the quality of this adaption that gives the final product a "timelessness" unaffected by changes in style or bias. We have titled this book **Environmental Quality by Design**; the title, by its own definition, implies "not by chance," for an attitude of laissez-faire easily conflicts with qualitative goals. Whether the goal is the total preservation of the natural environment or the management of regional water resources or the building of a new shopping center, a conscientious effort must be expended to maintain a course of deliberate action. Although environmental quality is not free, society cannot be expected to bankroll a synthetic environment because the costs, including the loss of quality, are much greater. Minimum standards exist for almost everything, including how close one may safely walk to the edge of a cliff. Minimum standards, although in theory are good, are usually antiquated, inflexible, and of little use other than as guidelines to prevent absolute disaster.

the urban region

The Design Process in Theory

South Florida is blessed with a rich and diverse environment, as delicate as it is bold. Its warm subtropical climate is attractive to residents and visitors alike; however, it is this same warm humid climate that maintains the environment at a threshold level susceptible to the impacts of radical change. The built and natural environments in southern Florida are so tightly joined that any decision that does not consider them as a single unit is destined for less than satisfactory results. Arthur R. Marshall (1972), an ecologist, succinctly defined the southern Florida environment as an organism consisting of interrelated land uses for the urban community and for natural processes. Although each of the two land uses depends or is affected by changes in the other the policies and plans that direct their management are done as if one or the other does not exist.

Within the built environment, associated environmental, social, economic, and political values are tightly woven as an ecosystem of man and nature. The matrix of this association includes as its major axis the natural system (land, air, water, space, etc.), public investment (government services, utilities, etc.), and private development (investment and development of consumptive products and services and markets that maintain a viable community).

As one can understand from the simple diagram in Figure 50, it is often too easy to specialize in one of the three areas of the ecosystem, addressing only the issues that are familiar or in some cases where investigations are required by law. Proper planning and design is not a simple job; if it appears to be so, one is usually not doing it properly. This book will by no means make the job easier in terms of time and commitment, but, assuming that one wants to design and plan properly, it offers advice as to what questions to ask and of whom to ask them. The southern Florida region is extensively studied, and there are many well-documented reports, essays, and data, numerous and thorough in their consideration of selected parts and factors of the regional ecosystem. We must swing back, however, from an era of specialization to an era of synthesis, that is, bringing it all together. There are areas about which we presently know very little that when properly investigated will possibly bring new insights. Moreover, we realistically can no longer continue to define problems simply for the sake of the art, but must move more positively toward their resolution. Specialization, while an important factor in the design process, too often acts as blinders to more creative thought. Had the engineers who carried the design and development of the internal combustion engine to its present high level of use thought to consider the external impacts of fuel consumption, noise, and air pollution, a much more efficient and "timeless" machine would have been developed. Specialization has its place on the assembly line, but not in the decision-making or policy-making segments of the total spectrum affecting environmental quality.

Defining the Urban Region

Part Two adequately defines and examines the ecological regions of southern Florida, identifies their constraints and opportunities, and recommends policy and guidelines for their preservation relative to development opportunities. The built environment is defined as those areas that are either constructed upon, or managed and controlled by land use regulations; it includes the interior and exterior spaces of the community as well as public and private lands. The urban region is that part of the southern Florida ecosystem in which the built environment exists—the community. To define the geographic limits of the urban region is not easy, however, for it is tightly integrated with the ecological regions upon which it depends

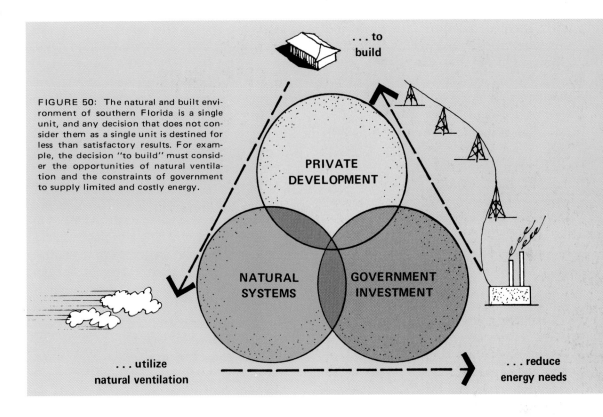

FIGURE 50: The natural and built environment of southern Florida is a single unit, and any decision that does not consider them as a single unit is destined for less than satisfactory results. For example, the decision "to build" must consider the opportunities of natural ventilation and the constraints of government to supply limited and costly energy.

... to build

PRIVATE DEVELOPMENT

NATURAL SYSTEMS

GOVERNMENT INVESTMENT

... utilize natural ventilation

... reduce energy needs

for resources and amenities. Most important, it cannot be defined as that area where the wilderness ends and the paving begins, for many land uses such as agriculture, parks, and water conservation areas, while not developed with brick and stone, are managed by an urban-type policy of control and production. All southern Florida must therefore be considered to be at some level of urbanization, for there is not one area that is not either developed, zoned, or managed to some degree.

Examination of Part One illustrates the width and breadth of the built environment in southern Florida. The urban region is all encompassing, and therefore the policies that will guide future renewal and development must address this entire region, its paved as well as unpaved areas.

Major conflicts within the urban region arise with incompatible land uses. For example, high-density development makes a recharge area less pervious, canals and wetlands are unable to assimi-

133

late excessive quantities of wastewater, excessive noise and air pollution from expressways penetrates the privacy of residential areas, etc. It is therefore easy to speculate that the control of incompatible land uses should be the primary concern for urban planning and development practices. However, the process must go beyond the distribution of land uses on a map; it must include the three-dimensional definition of tolerances between people, their needs, and the various uses of the land. Urban planning and design, that is, the art and science of directing the orderly growth and development of a community, has as its basic goal to improve the overall quality of the environment. Private development, on the other hand, tends to limit its consideration to the extent of the property line, seeking to maintain a margin of profit and depending upon government to satisfy the community perspective. The environmental quality of the urban region as a place in which to live is affected less by the direct involvement of the planning institutions, however, and more by the side-effects or by-products of our actions which may cause excessive crowding, noise, unemployment, increased municipal costs, etc. It is these by-products of building which this section will address as a means of improving the quality of the built environment.

Society's habitat is basically an urban one. As an urban species—gregarious, enterprising, and social—mankind's greatest works are in the city, although he seeks the solitude and beauty that nature represents (Becker, 1974). The quiet single-family neighborhood or the insulated and private condominium apartment have become the popular abode. One does not "live" in the city, but since the city provides work, entertainment, and protection, one cannot live without it.

"How has this fantastic reversal of earlier expectations come about? After all, in the past, urban living was thought of in terms of 'urbanity.' The city has been the place where a man went to live the better life. Many of the great images of our poetry have the city as the center of man's cultural and civic existence. Athens and Rome reverberate in our memories. Even more profound is the image of Jerusalem—Jerusalem, the spiritual city, foreshadowing the heavenly Jerusalem that is to come. And when the prophets spoke of their heavenly visions, they were visions of a city with gates of amethyst and streets of pearl. The city gave the promise of paradise. It was the rural peoples who were the barbarians, living beyond reason and grace, in the primitive countryside. Now the roles are reversed. Life and beauty are in the wilderness, and we have a doom-laden sense that our urban order is overwhelming us" (Barbara Ward, **An Urban Planet**, 11th Annual Gropius Lecture, Harvard University).

Urban environments become degraded and unpleasant not from too many people, but from the incapacity of necessary systems to service the needs of the people—water, air, roads, waste water treatment, housing, parks, jobs, etc. If demands of the people were balanced with the capacity of the system, environmental quality would be achieved (Veri, 1974). The best advice one can receive when evaluating the environment is to think at the broadest scale; to limit consideration only to the extent of the property bounds of a new apartment building or the right-of-way of a new road is an error that has overshadowed good judgment. To varying degrees, each segment of our habitat depends on interaction and support from all the other segments, most of which are located beyond the limits of the property line and which, in fact, define the total environment since man moves and lives in a multidimensional space (visual, audible, olfactory, etc.).

To achieve a proper balance of qualitative land use, the constraints and opportunities of any area or service within a community must be examined relative to its area of impact: physical, social, economic, and ecological. It is not until the extent and magnitude of the demands upon the natural and built systems are adequately resolved that proposals can properly be advanced or alternatives evaluated. For example, when land use densities are increased to provide more homes or new roads are built to move more people, the original problem of supply may not be

eliminated, nor even decreased, but multiplied. Only a certain number of cars can effectively operate on a road at one time, and the addition of 1,000 cars more per hour to a road already being used to capacity creates many problems: sixty-mile per hour traffic may drop to 20 miles per hour, idling cars generate many times more toxic fumes than cars traveling at 40 miles per hour, a 10-minute trip turns into an hour trip, noise increases, tempers flare, and accidents increase. Similarly, when parks become overcrowded, polluted, or the waterfront is privately owned and blocked from view, the public seeks the pleasures which once existed nearby but now are remote. Roads beget roads, and now it is necessary to travel to fish or swim in yet "undeveloped" areas. Whatever the scale, the objective is to avoid overloading environmental tolerances beyond their capacities. Government has the responsibility to provide services as well as direct development to within the tolerances of these services. Private industry has the responsibility to be aware of these tolerances and to make decisions relative to them.

Compared to past societies with fairy-talelike, quaint communities, the modern community is sometimes viewed as a physical and social failure. There is an occasional cry to return to the ways of the past, to look to our ancestors and their simple values to direct our decision making. However, it is not that the values of

modern societies have changed from those of the past, only that single-purpose technology has developed to produce more pollution than can adequately be tolerated by existing facilities or resolved by outmoded planning techniques. For example, the use of supersonic aircraft before all associated noise problems are solved clearly illustrates the effects of single-purpose production and technological by-products that exceed acceptable community tolerances. Pollution occurs whenever demands are greater than capacities: visual pollution when our eyes are confused; water pollution when wastes exceed the assimilation capacity of natural water bodies; poverty pollution when one cannot get a job or afford to live in the community. Technological or political quick-fix solutions in terms of short-term pollution abatement or welfare subsidies merely push the problem to another part of the urban system. New York City, one of the nation's wealthiest cities and major financial capital of the world, is facing bankruptcy trying to subsidize one portion of its environment at the expense of another. Many quick-fix schemes are cosmetic at best and therefore short-lived.

Modern societies are mere expressions of inherent goals shared by all humans from the beginning of time: protection, security, food, and shelter. The capacity for growth and development of modern societies, which are able to produce and consume at accelerated rates, may be limited more by the lack of space for the disposal of society's wastes than by the limits of land and resources for the production and consumption by the community (Environmental Protection Agency, 1973). The levels of noise, air, and water pollution, inconvenience, and fear which most urban dwellers encounter provide more insight for creative development practices than the standard front-end policies of restrictive zoning. This latter measure simply produces monotony rather than resolution of the issues. With a goal of enlightened policies in mind, public and private interests have an inherent responsibility to work together, to develop the methods and means for the creation of a quality environment. One major drawback of our present design process is its reliance upon outmoded methods to direct and utilize modern technology. Restrictive ordinances stymie imagination rather than allowing one to plan as rapidly as he can build. Performance standards, that is, specifying what is to be achieved and not how, would greatly accelerate achievement of environmental quality.

the develop- ment process

Every development, whether public or private, begins with an idea to be programmed, financed, planned, designed, built, and utilized. Whether public or private interests are involved, the approach for realigning the project—the development process—follows a similar series of steps. The process shares many common characteristics for all projects; it is initially very general in concept with more detailed information and work occurring as the development moves toward completion. This pyramid-shaped process diagrammatically illustrates the work involvement and is inversely proportional to the risks if the initial steps follow a comprehensive order of priorities to shape and mold the final product to fit its environs. As previously defined, the environment consists of all the related factors that influence one's life: social, economic, political, and ecological. Too often the potential developer in his haste to build will eliminate the initial and very important parts of the process, and like trying to force a square peg into a round hole his project will face delay or denial. Many factors shape and give final form to the project including finances, zoning, social needs, ecological conditions of the site, etc. This section will discuss, and admittedly promote, development guidelines to avoid the conflicts and delays which most poorly conceived and executed developments encounter. These guidelines encourage development practices that are necessary to achieve environmental quality in the built environment.

Site Selection and Evaluation

The most critical stage of the development process is the site selection, and evaluation for this stage determines the potential and the constraints of the site. This initial step theoretically should establish the type of development and its intensity to be refined throughout the project planning process. Usually the potential developer does not have the luxury of choosing an ideal location, but must rely upon what is available at a price that he believes can return a reasonable profit. Many developments, creatively designed and financed on paper, have been delayed or even denied because they did not fit the site or location. Whether one begins with a site

TABLE 8

BALANCING ENVIRONMENTAL AND ECONOMIC COSTS AND BENEFITS

DECISION-MAKING INSTITUTION	QUESTIONS TO SATISFY
Developer	Market feasibility
	Land and development costs
	Scheduling
	Cash flow
	Delays
	Materials availability and supply
	Marketing and promotion
	and others
Government	Zoning
	Environmental impact
	Community amenities
	Providing timely services and utilities
	Tax gain/loss
	Building codes
	and others
Community	Change in character
	Tax burden
	Visual characteristics
	Loss of amenities
	Social compatibility
	Job resource
	and others

NOTE: Each sector is primarily concerned with maximizing the benefits while minimizing the costs, a balance that can only be achieved if each institution is considered separately. If, however, all the questions are reviewed collectively, the task of satisfying each to the benefit of its institution and to the satisfaction of the other institutions becomes complex. Certainly, trade-offs will be made between the groups to reach an optimum situation, i.e., that point at which most benefits are realized at least cost to the mutual satisfaction of all parties concerned.

Providing guidelines for the achievement of this optimum situation is the primary objective of this book. Although it is recognized that not all the land area of South Florida should be developed in equal intensity—and that some areas should not be developed at all—suitable areas must be identified and rational alternative methods for building and development be followed.

or an idea, or both, the physical characteristics of the site and its location in the community will determine the feasibility of the idea.

Site selection and evaluation must consider two important factors in the initial planning process (Tables 8 and 9): (1) the opportunities and constraints of the site in relation to the proposed development, such as elevation, hazards, soil types, water table elevation, drainage characteristics, etc.; and (2) the impact upon the community at large and especially the adjacent neighborhood, inasmuch as the impact will vary with the size of the development and its intensity.

During the initial feasibility study, three significant factors to be investigated are the market, the holding capacity of the site, and the receptivity of the community and local government to the proposed type of development as it fits into the overall master plan of the community. In establishing the market feasibility of the development, one must determine the overall demand, the land use mix, and the

absorption rate of the market. This work, usually performed by market experts, all too often is the extent of site evaluation. The site must also be analyzed in terms of its ecological characteristics, its physical holding capacity, the impact of surrounding land uses and future developments, and such special amenities of the site as vegetation cover, views, bodies of water on or near the site, access to the site, and availability of key utilities to the development, all of which can contribute significant costs or benefits to the overall project. An ecological analysis should involve the input and advice of related natural scientists whose experience and understanding of the area will quickly reveal hidden opportunities as well as constraints in terms of land use and project design. The natural scientist can act as an integral part of the development process and eliminate risks and costs throughout the design and planning stages.

Many variables and their priority must be identified and evaluated in terms of proper development practices. When these factors have been evaluated, the developer must consider the profit plan, determine the objective of the development, and decide if it is worth the time, energy, and risk involved. Only after all of these factors have been considered can an intelli-

FIGURE 51: The development process usually begins with an idea and ends with implementation. The process becomes more complex and more detailed as implementation is approached.

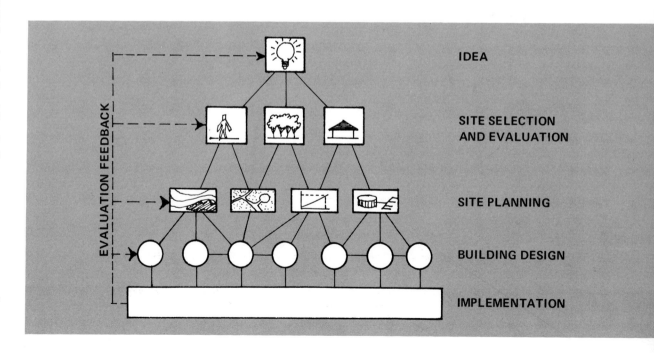

gent decision to proceed be made. Every project, site, and developer is unique to some degree: however, the basis of one's decision making must be made relative to the environmental setting of the proposed development, both in time and space. It is impossible to propose within the limited space available every design detail by which proper development practices can be accomplished. Certain rules and principles of planning and decision making are recommended, however, to illustrate the process by which a comprehensive approach can be achieved.

Site Planning

The land uses are decided and the utilities are arranged on the site in this stage. Site location and the type of development will, by necessity, be the primary concerns of most developers. However, the amenities and physical characteristics of the site will to a large degree indicate the optimum layout. Site planning decisions are critical to the overall environmental quality of the development project, for it is in this phase that the juxtaposition of various land uses is decided. The orderly development of any project must be preceded by the orderly analysis and arrangement of the land uses upon the site.

Any excessive noise that permeates an area affects the amenity and therefore the land use. A freeway which causes high noise levels on adjoining properties, influences the type of land use and the design and acoustical insulation of buildings. Vehicular traffic is the foremost noise and air contaminant in the southern Florida region (Veri, Peterson, Gerrish, 1973), and the disturbing noise of an improperly muffled truck, car, or motorcycle offends the ears of the largest number of the urban population. In addition to the close-range effects of noisy vehicles, the background "roar" of heavily traveled freeways creates corridors of background noise over and above the ambient sounds of the community. Unlike other background noises, the noise of an expressway is most noticeable at night when the rest of the community is quiet. Choosing a site within an area where noise is generated will require more expensive building materials if a given level of environmental quality is to be achieved. It is callous, to say the least, to permit new residential developments to the right-of-way lines of major roads and expressways. Obviously, neither the developer nor the planner considered the impact of adjacent land uses on the new homeowners. It is not by accident that most older developments located near areas of poor amenities and pollution

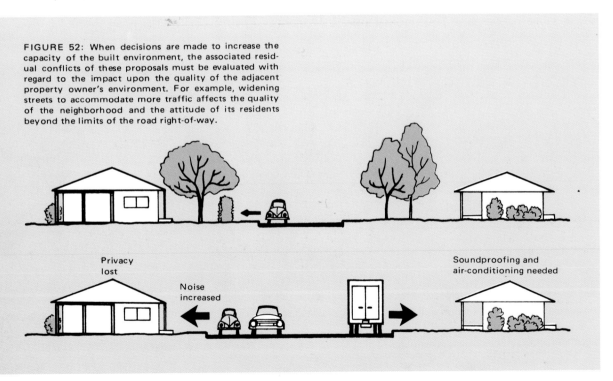

FIGURE 52: When decisions are made to increase the capacity of the built environment, the associated residual conflicts of these proposals must be evaluated with regard to the impact upon the quality of the adjacent property owner's environment. For example, widening streets to accommodate more traffic affects the quality of the neighborhood and the attitude of its residents beyond the limits of the road right-of-way.

Privacy lost

Noise increased

Soundproofing and air-conditioning needed

140

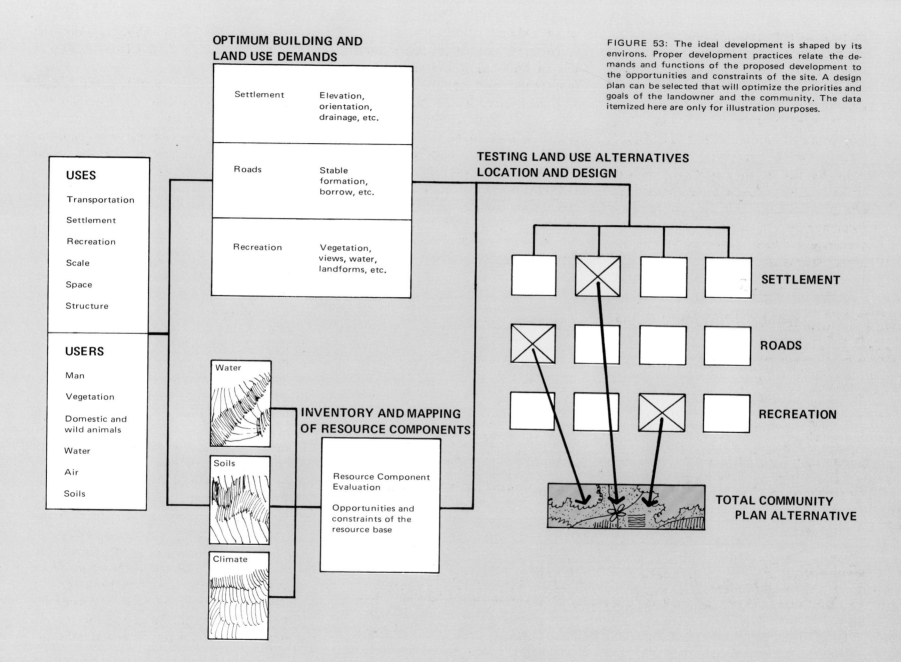

OPTIMUM BUILDING AND LAND USE DEMANDS

Settlement	Elevation, orientation, drainage, etc.
Roads	Stable formation, borrow, etc.
Recreation	Vegetation, views, water, landforms, etc.

FIGURE 53: The ideal development is shaped by its environs. Proper development practices relate the demands and functions of the proposed development to the opportunities and constraints of the site. A design plan can be selected that will optimize the priorities and goals of the landowner and the community. The data itemized here are only for illustration purposes.

USES

Transportation

Settlement

Recreation

Scale

Space

Structure

USERS

Man

Vegetation

Domestic and wild animals

Water

Air

Soils

Water

Soils

Climate

INVENTORY AND MAPPING OF RESOURCE COMPONENTS

Resource Component Evaluation

Opportunities and constraints of the resource base

TESTING LAND USE ALTERNATIVES LOCATION AND DESIGN

SETTLEMENT

ROADS

RECREATION

TOTAL COMMUNITY PLAN ALTERNATIVE

TABLE 9
THE DEVELOPMENT PROCESS

1 INITIAL FEASIBILITY

The three significant factors to be investigated are:

A. Market:
- demand
- mix of land uses
- absorption rate
- required amenities in order to be competitive

B. Site:
- ecological characteristics
- land area holding capacity in terms of proposed uses
- impact of surrounding land uses
- relation to surrounding ecology
- special amenity opportunities (view, vegetation, bodies of water, recreation, access, etc., on or near site)
- access to site in relation to service area
- availability of key utilities essential to development

C. Government:
- determine agencies within government to contact
- determine attitude of local government and community toward growth and toward type of development proposed
- understand local zoning and building code requirements
- determine utilities and pollution control requirements of EPA
- measure costs and benefits to community in relation to public services and facilities

Evaluate above in relation to the **profit plan** and decide if it is worth time, energy, and risk involved. Only after this can an intelligent decision be made as to whether or not to proceed to the next, more detailed design stage.

2 THE GENERAL PLAN

The four significant factors to be investigated are:

A. Preparation of general plan:
- site plan layout
- building types
- utilities
- roadways
- character and appearance concept

B. Construction phasing:
- determine construction phasing in terms of market (private), need (public), and funding
- prepare construction phasing plan with completion dates for each
- prepare development costs schedule
- prepare critical path for decision-making approval and development

C. Environmental impact assessment:
- determine impact as part of planning evaluation
- federal and state requirements
- local ordinance requirements

D. Negotiations and review of overall concept:
- federal, state, and local public agencies review concerning:
  - zoning and plan development requirements
  - utility approvals
  - public hearings
- negotiations for financial approval

Evaluate the impact of the proposed development in relation to the area affected. Environmental assessment is a key tool to this process since it will identify needs and requirements for environmental protection and related costs.

TABLE 9

THE DEVELOPMENT PROCESS (Cont'd.)

3 DETAIL DESIGN OF FIRST PHASE

The significant factors to be investigated are:

A. **Analysis of particular aspects of first phase of project**
- market
- site
- government and community receptivity
- regional considerations
- economic analysis
- ecological impact
- planning considerations

B. **Detailed site planning and design**
- site grading
- utility layouts
- building prototypes—orientation, arrangement, floor plans, etc.
- amenity design—landscape plans, views, street furniture, etc.
- recreation facilities
- density limits within general plan
- determine exact type, number, and location of buildings, utilities, etc.
- community urban design—architectural, landscape, circulation, etc.

C. **Construction cost budgets**

D. **Sales or rental objectives**

4 FINAL GOVERNMENTAL APPROVAL OF FIRST PHASE

A. **Federal, state, local agencies review for approval**
- plan modifications as necessary
- revaluation of economics

B. **Preparation of covenants, etc.**

5 IMPLEMENTATION

A. **Final construction documents**
- site design, details, etc.
- working drawings

B. **Construction schedule**
- phasing
- administration

C. **Pollution abatement during and after construction**
- noise
- air
- water

This planning procedure is repeated for each phase of development. Each phase must be evaluated in terms of sales response, user response, cost, and construction problems. Continuous evaluation of environmental-engineering-aesthetics efficiency should be made. Within the guidelines established in the general plan, any additional phase can be adjusted to better reflect the current needs and desires of the users and surrounding community.

are inhabited by lower income families; most people who can afford to move will move away.

Site planning must consider the affects of the proposed development upon on-site amenities and adjacent land uses. Storm water runoff and drainage plans are usually engineered to achieve rapid disposal of excess water. Urban runoff contains many contaminants equal to secondary treated and sometimes raw sewage (Bishop, 1973). Depending on the location, the soils, and the percolation rate, storm water systems within the project area offer potential amenities for the development. The most successful projects are those that are able to retain runoff on the site for percolation into the groundwater or for the filtration and slow release through vegetative swales. The design and layout of such a system, however, will vary throughout the southern region. The principle to be followed is to look to the natural system for design and layout criteria. On the smaller site of 1 to 5 acres obviously it is difficult to achieve the same integrity as on the larger tracts of new communities. Government should establish the land use characteristics and performance standards to which the smaller developer can address himself.

Typical detailed site planning and design includes plans for site grading and planting, utility layouts, building prototypes, amenities, budgets for construction costs, and sales or rental objectives. A

proper layout of land uses must also apply innovative design and planning to fit the project to its environs, including water treatment demands, waste water and solid waste disposal, etc. Many of the maintenance by-products can easily be recycled: lawn clippings can be used for compost, storm water can be directed to irrigation and groundwater recharge, etc.

The accompanying tables and figures suggest many ways in which a site may be developed to create a high level of environmental quality. By creative planning, the impacts of development schemes on the community can be minimized.

The population and density of most developments are the principal factors of negotiation and objection by the community. Zoning does not always guarantee the developer the right to utilize the site as the building and zoning ordinance may allow. Since many community master plans and zoning maps were conceived by less than comprehensive means, and at best are general in their presentation since they are only guides and not prerequisites to decision making, the developer should not rely entirely upon them. Most communities interpret good planning as low density and rightly or wrongly have perpetuated the urban sprawl. Total population of the project should be the only criterion of evaluation because no matter how many units are placed per acre it is the accumulated magnitude of population with which the community must deal. Density distri-

bution on the site should be determined by the developer. He can best decide the type of housing or industry that best meets the market demands, satisfies the adjacent land uses, and meets the physical constraints of the site. No greater insult to land use planning has been made than the use of density as a criterion for environmental judgment. Once the total population is decided, its location and distribution on the site in terms of the exact number, type, and location of facilities and dwelling units should be determined by

the site characteristics. Density is a relative number that describes only the level of land use intensity; it is the total population on a tract of land (whether a single lot or a square mile) that will determine the demands and subsequent impact upon the services and facilities of a community. Density and distribution of facilities on the land will determine how the opportunities and constraints of the social, economic, and ecological values of the site are satisfied.

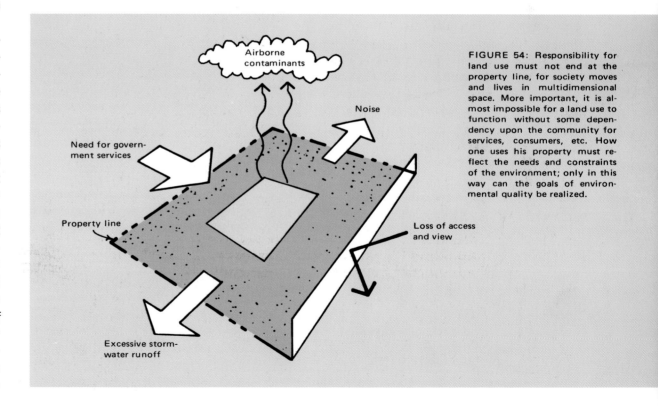

FIGURE 54: Responsibility for land use must not end at the property line, for society moves and lives in multidimensional space. More important, it is almost impossible for a land use to function without some dependency upon the community for services, consumers, etc. How one uses his property must reflect the needs and constraints of the environment; only in this way can the goals of environmental quality be realized.

Earl Starnes, Architect

Yiannis Antoniadis, Architect

Ted Baker, Landscape Architect

Building Design

Very few developments in southern Florida have utilized the inherent qualities of this region in the design and construction of buildings. The advent of the air conditioner enabled the creation of hermetically sealed environments to shut out the natural ventilation, and, by necessity, the increasing noise of the southern Florida environment. Recent studies, however, show that the poorly designed buildings in this region and, indeed, throughout the nation have increased energy consumption enormously. Air-conditioning and water heating account for over 50 percent of domestic energy use.

Very few developments have attempted to utilize the site or its location in the design and planning of buildings. Usually the site is altered by clearing, filling, and draining. The visual monotony of the southern Florida community reflects the lost opportunities of building to suit the site. For most of the year natural ventilation could be utilized for comfort control; solar energy can heat our water for bathing, laundry, and swimming pools.

The juxtaposition of buildings creates identifiable microclimates that add a dimension to the total environment. Parking lots located on the south and west sides of

high-rise buildings do not take advantage of the shadow cast by the structure, and these exposed asphalt lots are almost twenty degrees hotter than those shaded by trees or buildings. Situating buildings and streets to capture and direct the prevailing breezes can increase market value. How many new developments advertise "real Florida living?" Rather, we see new concepts such as California-style or Mayan-style architecture touted. The marketing potential of southern Florida usually ends with a picture of the sun with a palm tree on the beach.

The potential for environmental quality

146

Adonay Bergamaschi, Architect

H. Carlton Decker, Architect

Hernando Acosta, Architect

has not been adequately explored in southern Florida. The urban region is a fact of life, and the vast majority of Floridians can be expected to live on what is a relatively small area of land in megalopolises tied together by roads and markets. What goals we set for ourselves will determine what we leave to future generations. Our attitude must become one of homogeneity of purpose, to realize that a synthetic environment, even when economically feasible, will deliver less benefits for its costs. Effective performance standards must be defined as to what we want to achieve with less dependency on

restrictive ordinances that tell how to build. With these standards the impacts of urbanization upon the natural and built environment may not be eliminated, but they will be reduced to bearable limits, for all change brings associated impacts. One would insist, however, that as technology advances, the sources of conflict should be eliminated. We must begin now to use our ingenuity to guarantee society's rights to an environment of the highest possible quality. To do this we must utilize the knowledge we now have and the values we now realize to be important.

The search for total environmental

quality is much the same as the search for perfection. It may be impossible to achieve absolute perfection, but every positive effort decreases the distance to the ultimate goal. Every proposal for the use of land should follow the policy of seeking perfection. As knowledge grows, as more insight and technology becomes available, our decisions will be better. The ultimate answer may never be fully resolved, but we can always move closer to balancing development with ecological constraints, population distribution with land and services, thus guaranteeing living space that is comfortable, safe, and pleasing.

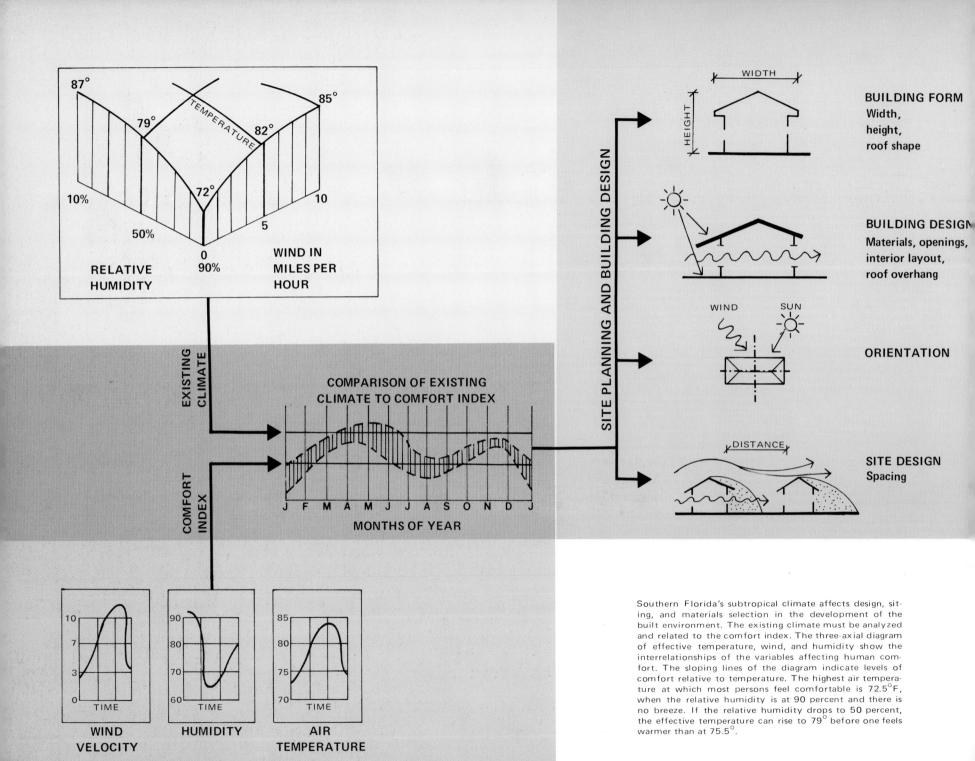

RELATIVE HUMIDITY

87° 79° 82° 85°
TEMPERATURE
72°
10% 50% 10
5
0
90%

WIND IN MILES PER HOUR

EXISTING CLIMATE

COMFORT INDEX

COMPARISON OF EXISTING CLIMATE TO COMFORT INDEX

J F M A M J J A S O N D J

MONTHS OF YEAR

WIND VELOCITY

HUMIDITY

AIR TEMPERATURE

SITE PLANNING AND BUILDING DESIGN

WIDTH
HEIGHT

BUILDING FORM
Width,
height,
roof shape

BUILDING DESIGN
Materials, openings,
interior layout,
roof overhang

WIND   SUN

ORIENTATION

DISTANCE

SITE DESIGN
Spacing

Southern Florida's subtropical climate affects design, siting, and materials selection in the development of the built environment. The existing climate must be analyzed and related to the comfort index. The three-axial diagram of effective temperature, wind, and humidity show the interrelationships of the variables affecting human comfort. The sloping lines of the diagram indicate levels of comfort relative to temperature. The highest air temperature at which most persons feel comfortable is 72.5°F, when the relative humidity is at 90 percent and there is no breeze. If the relative humidity drops to 50 percent, the effective temperature can rise to 79° before one feels warmer than at 75.5°.

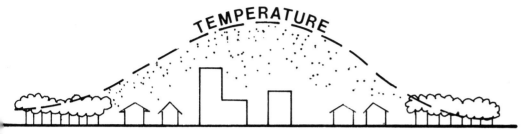

TEMPERATURE

Microclimate will vary with the intensity of land use. Asphalt and concrete will absorb and radiate more solar energy than the vegetated areas they replace. Additional heat sources include air-conditioning equipment and the operation of motor vehicles. The effect of these new heat sources will be manifested in an increase in air temperature as great as 5°C, creating a heat island. The heat island effect will create an upward movement of air, and a slight increase in cloud formation can be expected during the daylight hours, with a corresponding decrease at night.

55 designing the microclimate

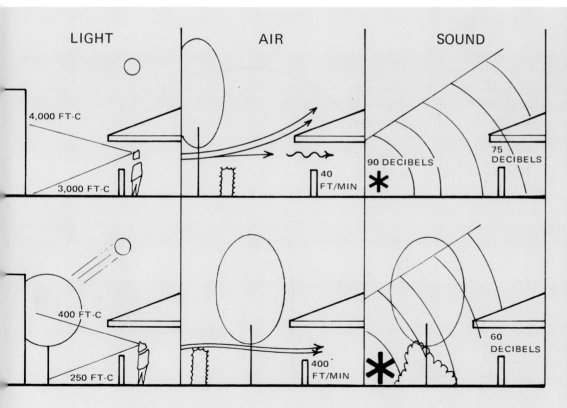

LIGHT   AIR   SOUND

4,000 FT-C

3,000 FT-C

400 FT-C

250 FT-C

40 FT/MIN

400 FT/MIN

90 DECIBELS

75 DECIBELS

60 DECIBELS

investigations by the Texas Engineering Station are the basis for these guidelines for providing human comfort. Interest is not limited to outdoor spaces alone, but includes the effect that site design has upon interior spaces as well. Planting can enhance a structure by creating aesthetic appeal and by modifying the microclimate.

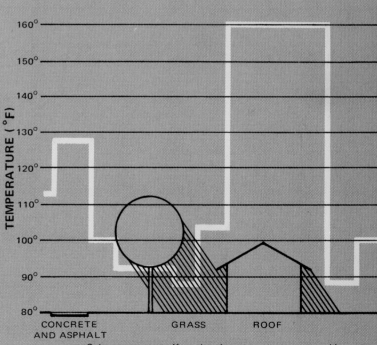

CONCRETE AND ASPHALT   GRASS   ROOF

Solar energy can affect the air temperature at ground level. Temperature on roof surfaces can reach twice the ground level temperature of shaded grass. Roof surfaces offer many opportunities for collecting solar energy for domestic uses.

water: a regional issue

The review of the water system in Parts One and Two provides a background for understanding the causes of existing water problems and of problems to be expected if present policies are continued. Southern Florida is a watery world; water defines and maintains the natural landscape. However, the present policies by which the resource is managed, and the practices by which the built environment is located and designed, conflict with the cycle of water resource renewal.

Three pressing issues concerning water management in this region are (1) the protection of the water supply, (2) disposal of waste water (sewage and runoff), and (3) preservation of a quality environment. It is a paradox that in this region where average rainfall amounts range from 45 to over 60 inches a year, water shortages occur that threaten to curtail future growth in some areas and the quality of life in others. Numerous alternatives have been proposed to increase the water supply of southern Florida; however, many experts agree that the problem results more from mismanagement than from lack of resource. Wasteful practices have greatly reduced an abundant supply.

A unique characteristic of the region is the critical value of the water system to the total environment. Water is recognized as the primary link throughout the entire natural and built environment, and its availability and quality is vital to the preservation of the richness and diversity as well as the very existence of these environments (Lugo, et al., 1971). In many cases, costly technology must be applied to bring the water up to adequate standards. The amount of usable water is directly related to its quality, and it is possible that the quality of water and the costs of purification may prove to be more of a limiting factor than total quantity. Furthermore, in some instances technology often replaces one set of problems with another, as evidenced by the possible threats to health resulting from excessive use of chlorine in water treatment.

Conflicting policies and actions tend to diminish the resource and result in less effective and more costly management procedures. In the urban region, per capita water use is increasing from an average 165 gallons per person per day 10 years ago to an average of 195 gallons today.

However, although per capita demand is increasing, the supply is diminishing as new urban development covers more land, reducing recharge and increasing water pollution over a larger area. Urban growth in areas where water is replenishable not only degrades the waters but also reduces the ultimate amount that can be used.

In Dade County alone water discharged to the sea through drainage canals equals five times the amount of water used by the populace annually. Urban runoff and treated waste water discharged directly into canals and other surface waters, bypassing wetlands with their beneficial cleansing action, results in lower quality in the receiving waters downstream. Concerned by the diminishing water resources needed for natural processes and conflicting land use policies, Craighead (1971, p. 46) aptly quotes Lamar Johnson, who in 1958 expressed with dismay the policy for water management in southern Florida: "After discussing the future demands for water in the Central and Southern Florida Flood Control District, he concluded that 'one drop of water must now preserve what two drops created.' "

It is not difficult to list the impacts resulting from conflicting courses of action. What is important is the urgent need for a comprehensive regionwide program of water resource management designed to cope with present and future problems of water supply, water quality, flooding, and with the preservation of a quality environment. South Floridians will probably never run out of water; it will undoubtedly be obtained from somewhere, no matter what the cost. However, although importing or reusing water has many obvious benefits, the costs must also be assessed. The best economic strategy is to make the existing system more efficient and to rely more heavily upon natural processes for maintaining quality, regulating flow, and providing storage. Continued abuse of this sensitive ecosystem will ultimately destroy it, a consequence that will force society to supplant natural processes with costly artificial systems that are not nearly as efficient or effective, and which, in terms of continuing energy problems, may not, in fact, be feasible. Conservation practices and water reuse may well increase the available supply; however, how much water may be made available by

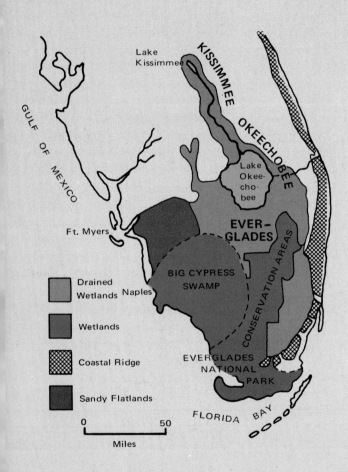

FIGURE 56: Location map of water resource areas of southern Florida.

more prudent management is difficult to answer. A study is urgently needed to determine the feasibility and the environmental-socio-economic impact of proposed alternatives, as yet untested in southern Florida.

Water management is a primary factor in decisions affecting land use, population distribution, and protection of natural resources in southern Florida. Changing priorities, new technology, and a better understanding of natural processes have made the limitations and deficiencies of the existing water management system painfully evident. Since water management can only be effective on a regional level, the first step in upgrading the system should be a regional evaluation of the strengths and weaknesses of the existing system. Only then should the system be modified or expanded. Furthermore, it would appear that this analysis and evaluation of the present system would be the least costly and most effective measure compared to the proposed alternative modifications and technology.

Environmental quality can result only from a concerted long-term effort and commitment to wise resource use, and not from unrelated decisions aimed at solving a particular and immediate problem. This chapter will recommend, on the basis of recent knowledge and technology, policies and programs for water resource conservation and reuse which could be considered once an overall evaluation of the existing system is made. Each recommendation will undoubtedly be affected by future technological advances, but anticipation of future breakthroughs should not limit the full use of natural processes and existing technology for water resource management (Veri, 1972; 1973).

The Regional Plan

Since water management can only be efficient on a regional level, the first step in upgrading the present water management system and governing policies should be a regional evaluation of the strengths and weaknesses of the existing system. Only then should it be expanded or modified.

Two primary reasons for the necessity of this evaluation are:

1. The present facilities of canals to provide a means of collecting runoff and impoundments to store water were designed essentially for flood control and drainage. More complete knowledge indicates that while these two objectives are being accomplished, the impact upon the cycle of water resource renewal results in a decrease in the optimum supply. Therefore, decisions regarding the District's facilities, land use, and management require a comprehensive understanding of the works in relation to the entire southern Florida environment.

2. Since the natural, urban, and agricultural water needs now compete for a relatively finite quantity of water, deci-

sion makers must discourage actions that may lead to long-term adverse environmental impact. Recently proposed management and technological modifications to the existing water management system are addressed primarily to specific issues of water management, such as increasing storage. Proposed mainly by engineers on a problem-solving mission, the methods and their results may not have been thoroughly studied. The feasibility and desirability of each proposal must be properly evaluated as well as its relation to the overall plan.

Although it is essential that water be managed more effectively, it is also essential that it be done in the context of a comprehensive planning process that provides for the resolution of other regional problems.

Comprehensive regional planning deals with the interrelationships between and among appropriate elements of the physical, social, and economic environment in order to optimize regional growth and change (South Florida Regional Planning Council, 1973). A thorough examination of water management proposals cannot be properly accomplished without considering the regional impact in relation to local characteristics and needs. More importantly, local requests conflicting with the overall comprehensive plan should not be approved.

The 1972 legislature passed several laws to improve the management of both de-

velopment and water throughout the state (see page 165). Under the Water Resources Act, the Central and Southern Florida Flood Control District (FCD), now called the South Florida Water Management District, was given new boundaries that extended their present authority in the Kissimmee-Lake Okeechobee-Everglades watershed to include the Big Cypress Swamp and Caloosahatchee River drainage basins. The FCD, under a delegated requirement of the Florida Department of Natural Resources, is now preparing a "water use plan" for this region. Concurrently, under the Act, the Florida Department of Pollution Control is required to prepare a "water quality plan." The combined plans will constitute the "state water plan" for this region. To ensure conformance with comprehensive regional goals and objectives, the state water plan should be considered a primary element. In this way, the regional planning and water management agencies can develop mutually supportive plans and programs that each can implement under its state-designated responsibilities.

Each major policy decision or program implementation should be examined as a Development of Regional Impact. Past decisions concerning water management have been made which have greatly affected the usefulness of the natural processes in the water basin. For example, the Kissimmee River was a meandering river with marsh areas that removed nutri-

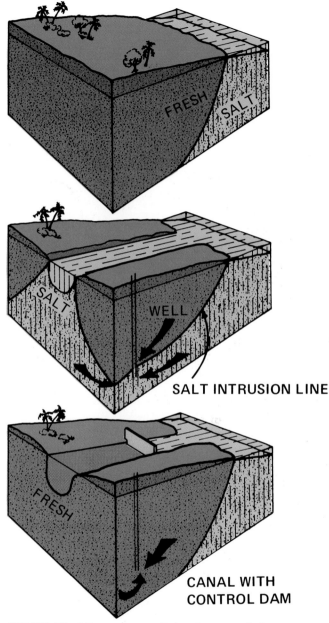

SALT INTRUSION LINE

CANAL WITH CONTROL DAM

FIGURE 57: Effects of controlled and uncontrolled coastal canals on seawater intrusion. (From: Sherwood, McCoy, and Galliher, 1973.)

153

ents, slowed water flow, and provided water storage. After the river was channeled and its extensive marsh areas drained, pollutants normally removed by the marshes now enter Lake Okeechobee. Moreover, while the benefits of the marshes are greatly reduced, the drained areas are used more extensively for agriculture, thus increasing the pollutant loading. Water control projects stimulate land development and emphasize the need for more comprehensive evaluation of all resource management proposals.

Land development in certain areas could easily conflict with natural renewal of the water resource. These areas include the presently undeveloped land along the east coast between the urbanized coastal area and the conservation areas, areas adjacent to and affecting the Big Cypress

Swamp water basin, and coastal marshes and swamps. The 1972 Environmental Land and Water Management Act requires the establishment of criteria to define "Developments of Regional Impact" and specifies the process by which regional planning councils review such developments. The same standards should be applied to public works projects involving any modification or expansion of the present system.

Subregion Alternatives

Studies are urgently needed to determine the feasibility and desirability of the many proposed alternatives for increasing the efficiency of water management. Let us emphasize again that the following possibilities represent current knowledge and technology. Each must be evaluated with

respect to the regional plan, to the impact upon other elements in the comprehensive plan, and to the related environmental constraints.

Federal regulations that require "dumping" excess water should be changed to allow conservation of southern Florida's water resource. Under natural conditions the southern Florida region experiences alternating cycles of flooding and drought. The ecosystem flora and fauna are adapted in a delicate balance to flooding during the rainy periods and fires during dry periods. Both periods are vital for sustaining and renewing the natural system (Douglas, 1945). In direct contrast to the natural cycle, the present management policy is designed to smooth these extremes. The present FCD system of canals, levees, control structures, naviga-

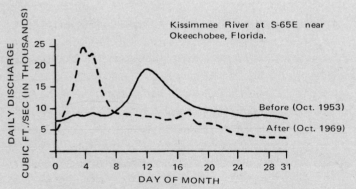

FIGURE 58: Example of change in rate of water flow due to canalization. The solid line is before and the dotted after. A forward shift in the sea flood. (Source: Marshall, 1972.)

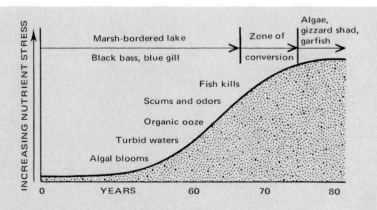

FIGURE 59: Theoretical curve of lake eutrophication. (Source: Marshall, 1972.)

tion locks, and pumping stations was superimposed upon an earlier network of drainage canals.

As a part of the system, the regulated storage areas of southern Florida are Lake Okeechobee, the conservation areas, and to a lesser extent, the canals. Numerous residential lakes and rock pits also impound water but are not considered to be a part of the regulated system.

Each of the four major storage areas is scheduled by federal law, which regulates the maximum quantity of water that may be stored at any given time. If, therefore, the water level in Lake Okeechobee or one of the conservation areas is above the prescribed water level, excess water is usually "dumped" (transferred by canals) seaward rather than into other storage areas. This policy is illogical when one considers the vital need for water within other parts of the system during dry periods. The present water storage capacities of the region could be used more effectively to reduce possible future shortages by transferring excess water to other downstream storage areas rather than wasting it by sending it to sea. Any storage modification must, however, consider the rate of flow, the quantity, and the quality of flow to such natural systems as estuaries and wetlands, since denying or reducing the present flow to these areas can have a major impact upon the ability of these systems to maintain their ecological balance.

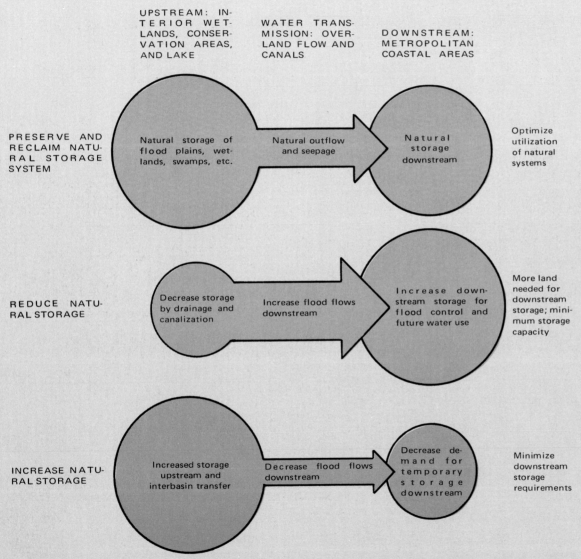

UPSTREAM: INTERIOR WETLANDS, CONSERVATION AREAS, AND LAKE

WATER TRANSMISSION: OVERLAND FLOW AND CANALS

DOWNSTREAM: METROPOLITAN COASTAL AREAS

PRESERVE AND RECLAIM NATURAL STORAGE SYSTEM

Natural storage of flood plains, wetlands, swamps, etc.

Natural outflow and seepage

Natural storage downstream

Optimize utilization of natural systems

REDUCE NATURAL STORAGE

Decrease storage by drainage and canalization

Increase flood flows downstream

Increase downstream storage for flood control and future water use

More land needed for downstream storage; minimum storage capacity

INCREASE NATURAL STORAGE

Increased storage upstream and interbasin transfer

Decrease flood flows downstream

Decrease demand for temporary storage downstream

Minimize downstream storage requirements

FIGURE 60: Three alternatives for storage of surface and groundwater in the southern Florida region.

156

Utilize the drawdown zone of well fields for underground storage of treated excess storm water. The space between the drawdown water level and the maximum water table surrounding well fields can be described as potential underground water storage areas that can function much the same as surface areas. Because injection of untreated water could easily contaminate the aquifer, the water stored in the drawdown must be treated before it can be stored, depending on the quality of the collected water. An average drop of the water table by one-half foot over a 50-square mile area would create storage for 1 billion gallons of water (Greeley, Hansen, and Connell, 1973). Well fields associated with drawdown areas would be bordered by major canals with artificially maintained water levels. A substantial portion of excess storm water, which would otherwise be discharged to sea, would be captured for storage in the drawdown portion of the aquifer around wells. The total storage volume would be an additional supply utilized during periods of low rainfall. For this reason, drawdown storage could only be utilized most effectively at inland rather than coastal sites. To maintain the freshwater head, the coastal wells, which are more susceptible to saltwater intrusion, would be pumped at a rate that would not create large area drawdown. Inland well fields used in the drawdown plan would generally be upstream of major urban areas and closer to the conservation areas which provide a primary source of recharge water. Rainfall in the drawdown area would be a minor additional recharge source.

Utilize backpumping of treated excess storm water to storage areas. Backpumping has a high potential for conserving water. This method retrieves water that would otherwise be discharged to sea by pumping it back into the storage areas. Before this alternative can be implemented, however, such factors as the capacity of available water storage areas, costs of implementation, and most important, **the quality of the water to be backpumped must be evaluated** (Ashley and Veri, 1974). Backpumping cannot be applied indiscriminately. The side effects of backpumping, such as the dewatering

of wetlands adjacent to the backpumping works, must be considered in determining if there will be a net gain in overall water supply. Furthermore, although artificial environmental conditions have already been imposed upon the conservation areas by the existing flood control works, backpumping may cause further environmental

alteration to these areas. A detailed environmental assessment must therefore be made before this alternative is undertaken.

Utilize the Floridan Aquifer for storage of treated waste water. The Floridan Aquifer underlies the region at depths averaging 1,000 feet. To use this aquifer

FIGURE 61: To dampen flooding during heavy rainfall, increase percolation, and decrease wasteful discharge to sea, a system of drainage retention ponds may be used to collect excess storm runoff in built areas. Although storm water drains directly into all the ponds, the smallest ponds provide the least retention. For rains up to perhaps 1 inch, each pond totally retains the runoff. For rains of greater intensity, water in the small ponds would overflow, via vegetative swales, into intermediate ponds. Prior to entering the intermediate pond, water is slowed and filtered to eliminate debris. The capacity of the intermediate ponds will be exceeded during extreme rainfalls, so overflows are directed to the large pond. Overflows from the large pond would be discharged into off-site areas, canals, etc. The retention pond design approximates the storm water retention capability of undeveloped land. Water within the pond systems will delay releases to off-site areas in a manner that closely approximates conveyance ability under natural conditions.

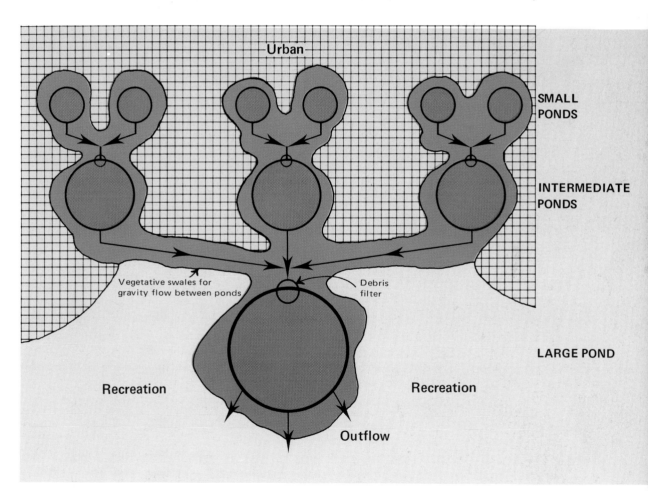

Urban

SMALL PONDS

INTERMEDIATE PONDS

Vegetative swales for gravity flow between ponds

Debris filter

LARGE POND

Recreation

Recreation

Outflow

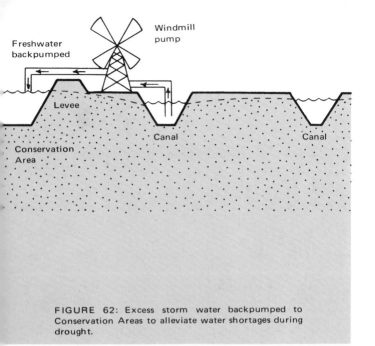

FIGURE 62: Excess storm water backpumped to Conservation Areas to alleviate water shortages during drought.

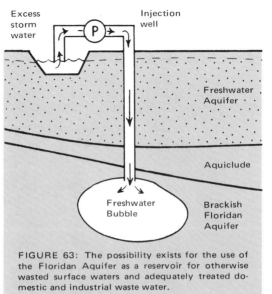

FIGURE 63: The possibility exists for the use of the Floridan Aquifer as a reservoir for otherwise wasted surface waters and adequately treated domestic and industrial waste water.

FIGURE 64: Simplified well field—aquifer storage concept. (Source: adapted from Greeley, Hansen, and Connell, 1973.)

as a potable water source would require desalinization, since the water contains chlorides in excess of 1,000 ppm. It has an even greater potential, however, as a reservoir for treated waste water (storm and sewerage). Most likely, untreated excess storm water will contain contaminants that will not be filtered out, and the retrieved water possibly could be more contaminated than waters available on the earth's surface. During wet periods, surplus water could be pumped into the upper portion of the Floridan strata, theoretically forming a "bubble" of freshwater floating on more dense saline water. Whether a freshwater bubble will form and remain intact, however, is still a theoretical concept. In times of drought these supplies could be withdrawn for treat-

ment and use. A guess is that at least 75 percent of the stored water could be recovered.

Recycle and reuse otherwise discarded previously used water. Reusing "waste" water has always been a vital process in natural systems. Recycling previously used water after treatment is imitative of the natural recycling and reuse that occurs in the hydrologic cycle (Figure 63). Today the volume of waste water in southern Florida amounts to almost 500 million gallons per day, and by the year 2000 this amount will increase to over 775 million gallons per day. These figures do not include the tremendous volumes of runoff from urban areas. The volume of domestic waste is estimated at 150 gallons per person per day although only 60 percent is

actually generated from residential usage; the balance comes from industries and infiltration of groundwater into sewers.

The major obstacles to the use of recycled water are public acceptance and the assurance that all pathogenic viruses will be eliminated (James, 1974). In addition to the obstacles, treatment costs may prohibit bringing this water up to potable quality for domestic use. More acceptable uses may include lawn irrigation, salt-front stabilization, and, after treatment, direct injection into the Floridan Aquifer for future recovery.

Utilize desalinization for recovering brackish water to relieve pressure on other freshwater sources. High-capital investment requirements, high-energy demands, and limited technology make this alternative a last resort compared with the other options. Energy demands for desalinization are extremely high. For example, it is estimated that to provide the present population of Dade County with enough desalted water to meet total present demand would require an energy source as large and as productive as the Turkey Point nuclear power plant operated by the Florida Power and Light Company. Furthermore, disposal of the hot, highly saline wastewater would require large cooling ponds. The damage done by direct disposal to the warm tropical sea would be extensive.

Revise present building standards and codes to encourage the use of water-conserving methods in all new construction and remodeling. Significant amounts of water are wasted by inefficient and deteriorated building products. Water-conserving plumbing fixtures should be required in all new construction and remodeling. For example, plumbing fixtures, including toilets, showerheads, and faucets, are available which reduce normal water use 30 percent to 50 percent compared with presently used designs. Brackish water from the Floridan Aquifer could be used for air-conditioning systems and power plants, if such were properly designed, instead of freshwater. Although limited, such conservation steps could relieve the demand on other freshwater sources and should be considered if the freshwater supply is restricted.

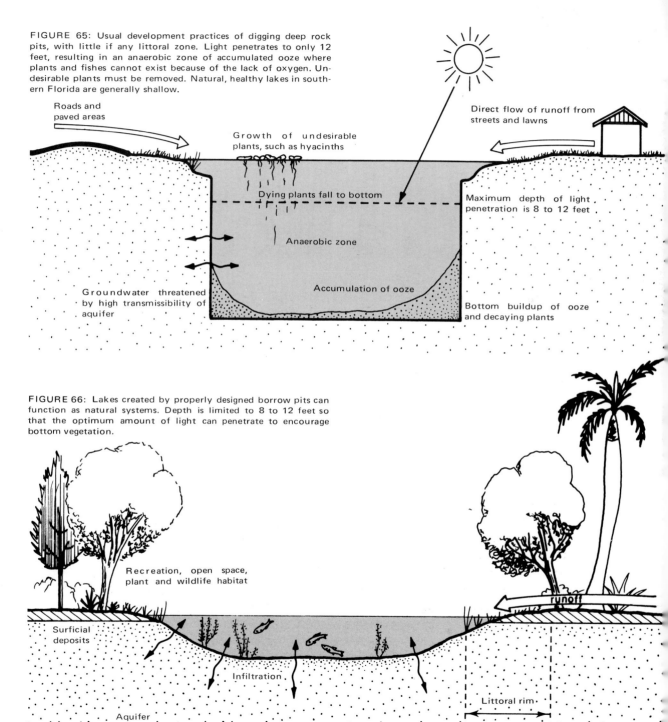

FIGURE 65: Usual development practices of digging deep rock pits, with little if any littoral zone. Light penetrates to only 12 feet, resulting in an anaerobic zone of accumulated ooze where plants and fishes cannot exist because of the lack of oxygen. Undesirable plants must be removed. Natural, healthy lakes in southern Florida are generally shallow.

Roads and paved areas

Direct flow of runoff from streets and lawns

Growth of undesirable plants, such as hyacinths

Dying plants fall to bottom

Maximum depth of light penetration is 8 to 12 feet

Anaerobic zone

Accumulation of ooze

Groundwater threatened by high transmissibility of aquifer

Bottom buildup of ooze and decaying plants

FIGURE 66: Lakes created by properly designed borrow pits can function as natural systems. Depth is limited to 8 to 12 feet so that the optimum amount of light can penetrate to encourage bottom vegetation.

Recreation, open space, plant and wildlife habitat

runoff

Surficial deposits

Infiltration

Aquifer

Littoral rim

Nutrient trap

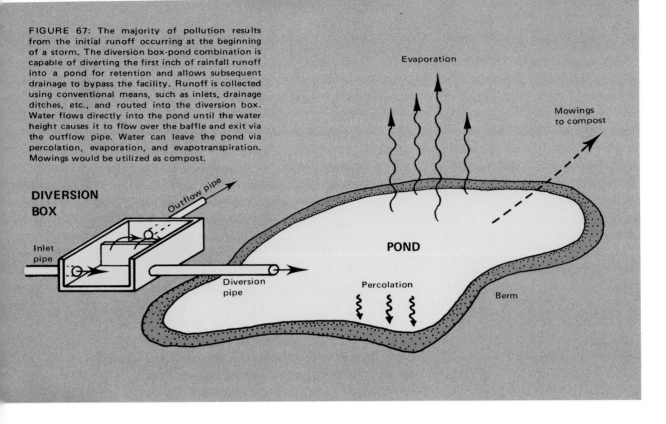

FIGURE 67: The majority of pollution results from the initial runoff occurring at the beginning of a storm. The diversion box-pond combination is capable of diverting the first inch of rainfall runoff into a pond for retention and allows subsequent drainage to bypass the facility. Runoff is collected using conventional means, such as inlets, drainage ditches, etc., and routed into the diversion box. Water flows directly into the pond until the water height causes it to flow over the baffle and exit via the outflow pipe. Water can leave the pond via percolation, evaporation, and evapotranspiration. Mowings would be utilized as compost.

DIVERSION BOX

Inlet pipe

Outflow pipe

Diversion pipe

Evaporation

Mowings to compost

POND

Percolation

Berm

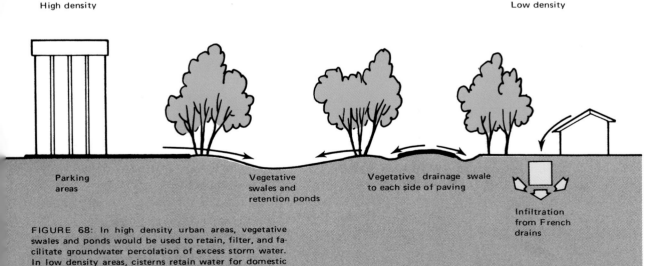

High density

Low density

Parking areas

Vegetative swales and retention ponds

Vegetative drainage swale to each side of paving

Infiltration from French drains

FIGURE 68: In high density urban areas, vegetative swales and ponds would be used to retain, filter, and facilitate groundwater percolation of excess storm water. In low density areas, cisterns retain water for domestic use.

Development Policy

It is possible for society to live within the environmental parameters of natural processes without destroying them. Significant changes in water quantity and quality are caused by development practices that disregard the constraints and ignore the opportunities provided by natural processes. The intense competition for natural resources, including space, especially in urban areas, has stimulated a dedicated revolution for the protection of our environment. This attitude has resulted in a new public awareness of the importance of our natural resources, especially water. This awakened consciousness ultimately requires an input of new types of data to urban planning. Traditionally, the urban planner has utilized hydrologic data only in terms of water supply, drainage, and waste disposal, although the natural processes of water resources renewal are equally important.

Water has historically played a key role in the design and policy governing urban growth and development. Land speculators and developers soon realized that property on a waterfront is highly marketable. Coastal areas in southern Florida were eagerly developed, not only because of their high and dry physiographic character, but more because of their proximity to the water's edge. High density usually exists in waterfront areas because of high land costs. As waterfront land becomes scarce and development is impeded by zoning or denials of permits to disturb submerged lands, inland drainage canals and real estate lakes become increasingly popular.

The landscape of southern Florida has been extensively altered by the clearing of land and the creation of impervious landscapes. Considered individually, modifications appear insignificant; however, collectively and over time, their impact is obvious. In addition to lowering water tables, rapid runoff into drainage canals changes the hydroperiod by reducing the length of time during which water is flowing, thus shortening the time for natural purification.

The rate of urban runoff varies with the percentage of land made impervious. In general, as urban density increases so does the amount of impervious surface and rate of runoff. At densities of four units per acre and average of about 20 percent of the surface is impervious; with 30 units per acre about 80 percent may be impervious. It is estimated that nearly half of the materials that cause pollution, which enter bodies of water, originate from storm water passing over and through urban and agricultural areas (Bishop, 1973). These additives result in a rapid acceleration of the aging (eutrophication) of these bodies of water. The changes in the hydroperiod and water quality in turn change the estuarine environment, which threatens the commercial and sport fishing and tourism. An optimum level exists, although not adequately defined, for using land without destroying the natural processes. Water is but one constraint and one opportunity in this process; however, it must be a major consideration in southern Florida to direct the distribution of population and the design of the built environment.

Each of the ecological regions described in Part Two is unique. Each deserves planning and design decisions that are equally unique. If it is agreed that indiscriminate clearing, draining, and paving can no longer be tolerated, then new practices must be developed and implemented. Part Two describes the opportunities and constraints of each of the regions, illustrates a simplified inventory of each region, and recommends development policies. It would be foolish indeed to suggest one or two specific methods that should be applied regionwide. It is, in fact, the application of standard building codes which has created much of the controversy we are seeking to eliminate. For example, while a local pollution control board will advise a builder to distribute water by swales and capture it for slow release in shallow ponds, the public works department will advise the same developer to collect excess storm water in concrete catch basins and discharge it immediately via pipes into the nearest canal or the sea. Only by responding to specific performance standards with innovation can the issue be adequately addressed. The builder has as much responsibility for defining innovation and preservation as does government, and it is only when both are working toward mutually shared goals that the conflicts will decrease.

South Floridians must decide what kind of environment they want to achieve, what sacrifices they are willing to make, and how much they are willing to pay to achieve these goals. Environmental quality is not free, but society can no longer expect to bankroll a synthetic environment dependent upon costly maintenance and repair. It is easy to say that real estate lakes are an amenity, but if their plan does not include proper design and maintenance guidelines then the facility becomes like an expensive piece of machinery that depreciates from the moment it is completed. On the other hand, natural systems grow, continually renew themselves, and require far less costly maintenance.

The value of maintaining a level of quality in surface waters is important beyond any aesthetic or recreational value. Degraded water quality results in higher treatment costs to the consumer, possible health hazards, delays in implementation of such water management alternatives as backpumping, and damage to the fishing and tourist industries.

TABLE 10
ENVIRONMENTAL CHECKLIST WORKSHEET: A Selected List

	Site Selection and Evaluation	Site Planning	Building Design	Construction	Management
A. AIR QUALITY AND CLIMATE					
Pollution					
Carbon monoxide	●	●	●		●
Hydrocarbons	●	●	●		●
Oxides of nitrogen	●	●	●		●
Sulfur oxides	●	●	●		●
Particulate matter	●	●	●	●	●
Odors and particles	●	●	●	●	●
Clarity and aesthetic quality		○	○		
Weather and climate (orientation, etc.)	●	●	●		○
B. WATER RESOURCES					
All types of water					
Freshwater availability	○	○			
Domestic water	●	●	●	○	●
Irrigation water	○	○	○		
Quality of water					
Physical	●	●	●	●	●
Chemical	●	●	●	●	●
Bacteria content	●	●	●	●	●
Aesthetics	●	●	●	○	●
Sewage treatment system	●	●	○	○	●
Water storage					
Drainage and runoff	○	●	○	●	●
Percolation and flood plains	●	●	○	●	●
Groundwater	●	●	○	○	●
Construction on aquifer recharge area	●	●	●	●	○

	Site Selection and Evaluation	Site Planning	Building Design	Construction	Management
C. GEOLOGY, TOPOGRAPHY, LAND FORMS, AND SOILS					
Soil and geologic stability	●	●	●	○	
Erosion and sedimentation	○	●	●	●	●
Aesthetic preservation	●	●	●	●	●
Pollution					
Pesticide	○	●	○	●	●
Herbicide		●	○	●	●
Toxic materials		●	○	●	●
Top soil removal		●		●	○
D. WILDLIFE AND HABITAT					
Habitats and migration areas	●	●	●	○	●
Specific wildlife	●	●	●	●	●
Fish contamination	○	●		●	●
E. OTHER RESOURCES					
Vegetation					
Destruction during construction	○	●	●	●	
Replacement valve	○	●	●	●	●
Scenic areas and parks	●	●	●	○	●
Wilderness areas and forests	●	●	○	○	●
Historic sites or structures	●	●	○	○	●
Archaeological sites	●	●	○	○	●
F. LAND USE (Adjacent)					
Conservation and preservation areas	●	●	●		
Recreational	●	●	●		

TABLE 10
ENVIRONMENTAL CHECKLIST WORKSHEET: A Selected List (Cont'd.)

	Site Selection and Evaluation	Site Planning	Building Design	Construction	Management
Agricultural	●	●	●		
Residential	●	●	●		
Commercial	●	●	●		
Industrial	●	●	●		
Institutional	●	●	●		
Division or disruption of an established community, e.g., dividing residential from recreational or disrupting planned development	●	●	●		
G. SERVICE SYSTEMS, FACILITIES					
Public schools	●	●	○		
Police facilities	●	●	○		
Fire protection facilities	●	●	●		
Health facilities	●	●			
Water systems	●	●	●	○	
Power systems	●	●		○	
Sewerage systems	●	●	●	○	
Refuse disposal systems	●	●	●	○	
Other public facilities and utilities	●	●			
H. TRANSPORTATION SYSTEMS					
Air transportation	●	●	●		
Land transportation	●	●			
Water transportation	●	●			
Effect upon air quality	●	●	●	○	●
Effect upon water quality	●	●	●	○	●

	Site Selection and Evaluation	Site Planning	Building Design	Construction	Management
Adequate access to site	●	●			
I. NOISE LEVELS AND VIBRATION					
On-site noise and vibration	●	●	●	●	●
Off-site noise and vibration	●	●	●		●
Construction in airport noise zone	●	●	●		●
Construction near high-volume highway traffic	●	●	●		●
J. SAFETY					
Fire safety	●	●	●	●	
Construction in flood plain	●	●	●	●	
Disease control; rodent control	●	●	○	●	●
Radiation and air pollution levels	●	●	●		●
Construction under high-voltage lines or over high-pressure gas lines	●	○	○	●	
K. THE ECONOMY OF THE AREA AND THE SURROUNDING REGION					
Tax revenues generated	●	●			
Inside the project area	●	●			
Outside the project area	●	●			○
Cost and availability of public services					
Inside the project area	●	●	●		
Outside the project area	●	●	●		

TABLE 10
ENVIRONMENTAL CHECKLIST WORKSHEET: A Selected List (Cont'd.)

	Site Selection and Evaluation	Site Planning	Building Design	Construction	Management		Site Selection and Evaluation	Site Planning	Building Design	Construction	Management
Employment				●	●	**Community structure**					
Location	●					Social structure of the community	●	●	●		
Accessibility	●					Relocation of people, buildings, services, etc.	●	●	○	○	○
Number of employees	●					Cultural attitudes	●	●	●		
Unemployment	●					Boundaries of political units or special districts	●	●	○		
L. SOCIAL REALITIES											
Population (Market Area)						**Employment opportunity**	●	●	●	●	○
Total population	●	●	●			**Educational opportunity**	●				
Density	○	○	●			**Physical, mental, and emotional health**	●	●	●		
Distribution	●	●	●			**Personal safety**	●	●	●	●	○
Ethnic composition	●	●	●								

● primary importance

○ secondary importance

land use and the law

It is proper, we believe, to end this book with a discussion of land and the laws that govern its use. Development of land and resources has been an ideology of the American society and economy since colonial days. The colonial period in America was one of great liberalization in terms of the right of the individual to bequeath his land as he chose. By the time of the Revolution, the concept of fee simple ownership was firmly established (Clawson, 1973). In its strictest interpretation, fee simple ownership gave the owner the right to the land from the center of the earth to the zenith of the sky to be used only as the owner saw fit. With this freedom to use land, however, came the need to regulate that use for the good of the community (Bosselman, Callies, Banta, 1973).

The colonies enforced strict land use regulations designed to promote health and safety. Regulations in urban areas resembled those of London. Some of the early regulations governing land use concerned setting aside land for agriculture, putting restrictions on building heights and materials, and separating land uses that were conflicting.

With the recent advent of environmental and social concerns, decisions affecting land use have shifted even more in favor of common interests. What was proper, or at least tolerable, for a landowner to do with his land when his neighbor was many miles away may have become intolerable when the neighbor is only a few feet away on the adjoining suburban lot. As we can see in southern Florida today, when populations and densities increase there is a greater reliance upon government to control and regulate land use, to provide roads and services, to manage water, and to control pollution and crime.

Government tends to express the values of its constituents, and the higher the level of government, the more diverse the constituency. At the local level, where decisions of local government tend to reflect individual profit and loss considerations, local jurisdictions cannot commit the greater part of their land and water resources to uses that are valued primarily by the larger society outside of the local arena. A community that must generate its own revenues for financing public services is not likely to commit space within its jurisdiction for social use.

In contrast, actions of national govern-

ment reflect the value structure of the nation at large and proceed from a power base not available to local governments. A nation of mobile people may reasonably expect that their interests and welfare should be protected in the use of local resources even in areas far from where they may live. National government can make decisions to leave certain areas as wilderness, regulate air and water quality, and affect similar national concerns. Traditionally, however, local governments have made most land use decisions regarding planning or zoning.

Most of us understandably lack a clear knowledge of how land use controls are administered and by whom. Although space does not allow us to discuss all the laws and regulations affecting South Florida, summaries of those concerning land use, resource allocation, and pollution control are presented. Similarly, the interaction of local, state, and federal agencies and their jurisdictions and responsibilities are itemized in written and chart form. This information hopefully will assist the reader in his decision making.

General Authority

Depending upon geographic location and the scale of development, land ownership and the use of resources upon and near it are regulated by levels of government charged by law with this responsibility. The basic powers exercised by the various levels of government are not special powers, but are part of the general governmental authority to deal with the economic, social, and natural environment. Federal and state levels derive their authority from their respective constitutions, and local government by delegation from the state. The land, water, and air of southern Florida are overseen by many levels of government agencies. Figure 69 is a generalized diagram showing the overlapping governmental jurisdictions for the land and water regions.

The federal government has jurisdiction over the territorial sea, the contiguous zone, and the continental shelf. States have property rights to the submerged land and jurisdiction over the water column out to three miles from shore. Recently Texas and Florida received nine miles out from shore in the Gulf of Mexico on the basis of historic claims. States may exercise control over their own citizens beyond the territorial sea. The constitutional authority of the federal government over all navigable water areas of the nation is based upon the so-called Commerce Clause of the United States Constitution. The most pertinent example of this authority is embodied in the regulatory functions of the Department of Transportation through the U.S. Coast Guard. Within the scope of this broad power the Coast Guard wields paramount authority over a wide range of activities and conditions related to local waters, both inland and coastal. A partial listing of these activities includes navigation, waterfront construction, water pollution abatement and prevention, restoration of eroded beaches, flood control, creation of wildlife preserves, and establishment of national park and offshore monuments for recreational use. Passed in 1953, The Outer Continental Shelf Lands Act establishes an exclusive leasing program by the federal government for areas of the continental shelf beyond state boundaries.

The Department of Defense through the Army Corps of Engineers controls all construction or maintenance activities occurring within navigable waters of the United States. Under their statutes, permits are required for these activities before work can be begun. The Department of Transportation through the Coast Guard enforces all federal laws on the high seas (beyond the states' territorial limits) and requires permits for construction of all bridges over federally maintained navigation channels.

At the state level, the Trustees of the Internal Improvement Trust Fund (TIITF) have duties similar to those of the Army Corps of Engineers with authority over and power to issue permits for all activities on state-owned submerged lands and in sovereign waters to the line of mean high water, including man-made bodies of water connecting to naturally navigable waters. The Department of Natural Resources Marine Patrol is, like the U.S. Coast Guard at the federal level,

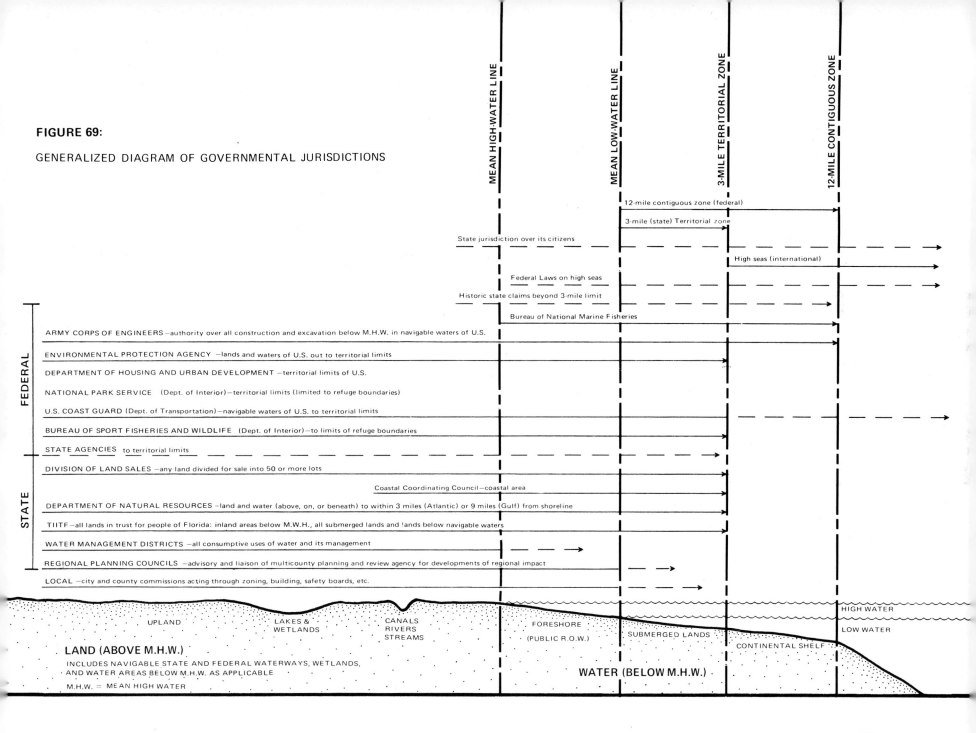

FIGURE 69:

GENERALIZED DIAGRAM OF GOVERNMENTAL JURISDICTIONS

MEAN HIGH-WATER LINE

MEAN LOW-WATER LINE

3-MILE TERRITORIAL ZONE

12-MILE CONTIGUOUS ZONE

12-mile contiguous zone (federal)

3-mile (state) Territorial zone

State jurisdiction over its citizens

High seas (international)

Federal Laws on high seas

Historic state claims beyond 3-mile limit

Bureau of National Marine Fisheries

FEDERAL

ARMY CORPS OF ENGINEERS —authority over all construction and excavation below M.H.W. in navigable waters of U.S.

ENVIRONMENTAL PROTECTION AGENCY —lands and waters of U.S. out to territorial limits

DEPARTMENT OF HOUSING AND URBAN DEVELOPMENT —territorial limits of U.S.

NATIONAL PARK SERVICE (Dept. of Interior)—territorial limits (limited to refuge boundaries)

U.S. COAST GUARD (Dept. of Transportation)—navigable waters of U.S. to territorial limits

BUREAU OF SPORT FISHERIES AND WILDLIFE (Dept. of Interior)—to limits of refuge boundaries

STATE AGENCIES to territorial limits

STATE

DIVISION OF LAND SALES —any land divided for sale into 50 or more lots

Coastal Coordinating Council—coastal area

DEPARTMENT OF NATURAL RESOURCES —land and water (above, on, or beneath) to within 3 miles (Atlantic) or 9 miles (Gulf) from shoreline

TIITF—all lands in trust for people of Florida: inland areas below M.W.H., all submerged lands and lands below navigable waters

WATER MANAGEMENT DISTRICTS —all consumptive uses of water and its management

REGIONAL PLANNING COUNCILS —advisory and liaison of multicounty planning and review agency for developments of regional impact

LOCAL —city and county commissions acting through zoning, building, safety boards, etc.

UPLAND

LAKES & WETLANDS

CANALS RIVERS STREAMS

FORESHORE (PUBLIC R.O.W.)

SUBMERGED LANDS

CONTINENTAL SHELF

HIGH WATER

LOW WATER

LAND (ABOVE M.H.W.)
INCLUDES NAVIGABLE STATE AND FEDERAL WATERWAYS, WETLANDS, AND WATER AREAS BELOW M.H.W. AS APPLICABLE

M.H.W. = MEAN HIGH WATER

WATER (BELOW M.H.W.)

TABLE 11

PERMITTING AND REVIEW BY FEDERAL, STATE, AND LOCAL AGENCIES

	MAJOR RESPONSIBILITY							PRIMARY JURISDICTION										
	Land	Water	Air	Fish and Wildlife	People	Services	Planning	Upland	Below M.H.W.	Navigable Water	Wetlands	Submerged Lands	Coastal	Areas of Critical Concern	Parks and Recreation	Preserves	Sea Bed	High Seas
FEDERAL																		
National Marine Fisheries Service [1]		●		●					●	●		●						
Army Corps of Engineers [2]		●					○		●	●		●	●				●	
Environmental Protection Agency	●	●	●		●		●		●	●			●					
Department of Housing and Urban Development	●								●				●					
Bureau of Sports Fisheries and Wildlife [3]	●	●		●					●	●		●	●		●	●		
National Park Service [3]	●	●		●			○		●	●		●	●		●	●		
U.S. Geologic Service [3]							○		●	●		●						
Bureau of Outdoor Recreation [3]	●	●													●	●	○	
Federal Highway Administration [4]	●				●		●		●	○			●					
U.S. Coast Guard [4]		●					●		●	●		●						●
STATE																		
Administrative Commission [a]	●	●		●			●		●	●		●	●					
Division of State Planning [a]	●	●	●	●	●		●	○	●	●		●	●		○	○		
Division of Florida Land Sales [b]	●							●										
Coastal Coordinating Council	●	●		○		○		○	●	●	○	●	●			○		
Department of Commerce	●	●		●	●	○	●	●	●						○			
Department of Natural Resources	●	●	●	●	○	●	○	●	●	●	●	●	●		○	○	○	
Department of Pollution Control	●	●	●		●	○		●	●	●	●	●	●			○		
Department of State	○	○	○															
Department of Transportation	●	●	●		●	●	●	●	○	○								
Public Service Commission						●												
Trustees Internal Improvement Trust Fund	●	●		○			○	○	●	●	○	●	●	○				
REGIONAL																		
FCD	●	●				○	●	●	●	○	●	○	○	●				
Regional Planning Councils	●	●	●	●	●	●	●	●	●	○	●	○	○					
LOCAL (within political boundaries)																		
Municipal Commission	●	●	●	●	●	●	●	●	●	●	●	●	●	●	●	●	○	○
Planning Department	●	●	●	○	●	●	●	●	●	○	●	●	●	●	●	●		
Building and Zoning	●	●	●		●	●	●	●	●	○	●	●	●	●	●	●		
Pollution Control	●	●	●	●	●	○	●	●	●		●	●	●	●	●	●		

ADAPTED FROM Shevin, 1973; Department of Natural Resources, 1973; Dickert and Domeny, 1974.

● Review and permit
○ Advisory
[1] Department of Commerce
[2] Department of Defense
[3] Department of Interior
[4] Department of Transportation
[a] Department of Administration
[b] Department of Business Regulation

---

charged with enforcing Florida law in the territorial waters of the state.

Federal and state authorities are the primary vehicle for environmental protection. At the federal level, such agencies as the Department of Commerce, the Environmental Protection Agency, and the Department of Interior have responsibilities covering the regulation, protection, and management of natural resources. At the state level, similar responsibilities are spread among a number of agencies; the most important are the Department of Natural Resources, the Department of Pollution Control, the Trustees of the Internal Improvement Trust Fund, the Game and Fresh Water Fish Commission, and the Department of Health and Rehabilitative Services.

Regional and local agencies focus attention on issues such as zoning, utilities control, public works, and the like related to land use usually above the mean high-water line. The respective local city or county commission, operating through various codes and ordinances, regulates such things as zoning, building, public works, public health, educational programs, etc. The Central and Southern Florida Flood Control District (FCD) is charged with responsibility for control and management of all competitive freshwater users in the region. The South Florida Regional Planning Council and the Southwest Florida Regional Planning Council undertake comprehensive planning (advisory) for southern Florida, including regional planning liaison with state and federal planning agencies, and they review and prepare comment on Developments of Regional Impact. Their jurisdiction includes the political boundaries of the respective county members.

The Review Process

Table 11 lists the various federal, state, and local agencies, their primary jurisdiction and major responsibility, and management reviews and permits necessary prior to undertaking particular land uses or site alterations. For simplicity, and since organization is constantly changing, this table is generalized and should be used for reference only. Its purpose is to illustrate governmental management jurisdiction in the use of lands and resources in public and private ownership. It is important to note that the state is currently considering the consolidation of all environmental agencies into one "super agency" headed by the Department of Natural Resources. The decision will be made after this book goes to press. While consolidation would provide "one-stop shopping" for developers, by shortening the time lag necessary for a multiagency review, it would greatly modify the checks and balances provided by multiagency review.·

Table 12 groups the several jurisdiction decisions important to southern Florida into two broad categories: upland and

TABLE 12
GOVERNMENT DECISION MAKING

GOVERNMENT LEVEL	UPLAND USE			COASTAL LAND USE	
	Routine Land Use	Land Use of Regional Impact	Environmentally Sensitive Land Use	Submerged Land Use	Marine Water Use
Local	●	●	●	○	○
Regional	○	●	●	●	●
State	○	●	●	●	●
Federal	○	○	●	●	●

● Need for clearance from designated authority level
○ Depending upon type of activity, clearance may not be needed

ADAPTED FROM Coastal Coordinating Council, 1974, page 130.

coastal. This table is adapted from a study of the Florida Keys conducted by the Coastal Coordinating Council (page 130, June 1974). "Upland uses" are defined as any proposed activity occurring above mean high water. Under this category, "routine use" is any human activity occurring well above the mean high-water line (MHW) in areas of low environmental sensitivity (i.e., activity above the hurricane flood zone, activity avoiding environmentally important vegetation, etc.). "Land use of regional impact" includes those activities of certain size or type that are defined as being of regional importance by the Florida Land and Water Management·Act. "Environmentally sensitive

land use" means proposed activities taking place in areas that are judged to be environmentally sensitive and productive because of their biophysical character. Coastal areas include the sovereign seas and upland areas having associated effect upon coastal resources. "Submerged land use" includes all modification or use of lands below mean high water. Specific agency responsibilities can be determined from Table 11.

Surrounded on three sides by water, southern Florida has no point greater than 70 miles from the sea. This proximity to the sea, coupled with southern Florida's relatively low terrain with extensive interior wetlands and the presence of navigable waters that bisect Florida's northern boundary (Lake Okeechobee via Caloosahatchee River Canal and the St. Lucie River Canal), make almost all the federal and state laws concerning water-oriented activities of major importance to land use decisions. The number of agencies involved in the approval and review of proposed developments generally increases with the scale, type of activity, and location. Furthermore, as development approaches the water's edge or wetlands, regulation greatly increases. Many estuaries and wetlands, environmentally and economically valuable (e.g., breeding grounds of fishes and sea critters), were irreparably destroyed by the process. These actions were, however, sanctioned by government, who with little substantial knowledge or jurisdiction was unable to foresee the impact of these precedents upon future decisions and policies of land use. It was these activities that spurred the nation, especially the coastal states, to delineate more clearly the use of water-oriented lands with respect to the social, economic, and environmental interests for the common good.

After considering private and public rights, one can see that private interests must be reconciled with concepts of sovereignty and public trust, and public concerns that overlap or compete in turn must be balanced against one another. The following sections summarize selected important laws relative to the three levels of government affecting land use and management decisions in southern Florida.

Federal Level

The federal government is limited by the U.S. Constitution to the exercise of those powers enumerated for it, primarily in Article 1, Section 8, which sets forth congressional authority. One of the broadest powers is the regulation of commerce among the states and foreign nations. Since one of the major uses in the coastal zone involves such commerce, and many conditions and actions can "affect" commerce, this power affords the federal government considerable leverage in the coastal zone. (U.S. Department of Interior, 1970, App. I).

The war powers of the federal government (U.S. Constitution, Art. 1, Sec. 8, also Art. 11, Sec. 2) are the legal justification for the involvement of the Army Corps of Engineers in works for improving navigation and for its authority to issue permits to construct docks, bridges, or other works that might obstruct navigation (U.S.C.A., Title 33). These powers are also the basis by which the corps regulates marine dumping of dredge spoil. The Environmental Protection Agency (EPA) has jurisdiction over marine dumping under Public Law (P.L.) 92-532.

The power and authority to manage federal property (U.S. Constitution, Art. 1, Sec. 8) is of great significance, both presently and in the future. As owner of fee title, or some lesser interest in real property, and as a government, the United States government can control the development or preservation of coastal areas held by it. (U.S. Department of Interior, 1970, App. H).

Virtually all states with shoreline margins, both coastal and otherwise, have adopted or are in the process of adopting comprehensive programs for reducing or eliminating air and water pollution. Because of public pressure, most county and municipal authorities are engaged in similar programs. These regulatory controls assume a wide variety of forms, the chief two of which are treatment of sewage and solid waste and discharge of pollutants from residential and industrial sources into both the air and bodies of water.

It is generally agreed that one of the main stumbling blocks against water quality improvement is the high cost of installing treatment plants and other sanitation facilities. Another problem, of course, is a lack of adequate enforcement. Where water pollution is derived from industrial plants and residential developments, however, a measure of control may be more easily exercised by requiring installation of antipollution devices at points of source and providing economic incentives for lower pollution levels. Significant effort is now being made in this direction by the federal government and an increasing number of states.

Florida, for example, has recently secured statewide voter approval for $200 million of state bonds for constructing waste water facilities, but this step at best only partially solves the problem. Antipollution controls, when not delegated to approved county units, are generally administered at the state level by the Department of Pollution Control and several affiliated agencies. In the main, this type of control appears to be the practice in most other states.

The most significant advances in pollution abatement policies are now taking place at the federal level. Legislative changes and additions to federal laws, as evidenced by the National Environmental Policy Act of 1969 and the Federal Water Pollution Control Act of 1972 (FWPCA), as amended, provide an impetus for comprehensive control. Under the amendments to the latter bill, broad powers with respect to establishment of water quality standards and enforcement procedures are vested in the Environmental Protection Agency. Section 201 of the amendment provides for facilities planning, and Section 208 for areawide waste-treatment planning. With some contributory participation on the part of coastal states, the federal government is committed to an $18 billion expenditure in the three-year period beginning fiscal year 1973 for installation of sewage treatment plants and other improvements for local waters. The Coastal Zone Management Act of 1972 partially funds and calls for cooperative planned use of coastal areas by the federal government and the several states.

In order to implement the Intergovernmental Cooperation Act of 1968 (P.L. 90-577, 82 Statute 1098), the Office of Management and Budget (OMB) issued Circular No. A-95 in which guidelines were given for comprehensive planning cooperation between all existing federal agencies and state and local planning agencies for proposals for federal grants-in-aid. Prior to this action, federal, state, and local programs had operated in an information vacuum, which often resulted in duplication of efforts or conflicts in policies or programs among agencies. Programs covered under Circular No. A-95 encompass most of the agencies that previously had been working independently in planning and development.

The OMB further suggested that the states designate an agency to receive and disseminate information to state and local agencies. In Florida, the State Planning and Development Clearinghouse (Department of Administration, Bureau of Planning) performs this function. Chapter 216.212 of the Florida Statutes conversely requires that every state agency must submit any request for federal assistance to the secretary of administration for review before submittal to federal authorities. One result of this increased exchange of information has been the realization that a positive national policy on the environment is necessary. (Still, 1972.)

In May 1969, President Nixon established the Cabinet-level Environmental Quality Council and charged it with taking a broad overview of environmental problems and proposing new approaches. Congress enacted two related measures: The National Environmental Policy Act of 1969 (NEPA) (P.L. 91-190) and the Environmental Quality Improvement Act of 1970 (EQIA) (P.L. 91-224). The council was then abolished, and the Council of Environmental Quality in the Executive Office was established by NEPA. The EQIA supplied the new council with expanded staff support in the newly created Office of Environmental Quality.

Probably the most far-reaching step in pollution abatement and control is the NEPA legislation. Section 102 of this act requires federal agencies in every recommendation or report on legislation or other major federal action that may significantly affect the quality of the environment to include a detailed statement on the environmental impact, any adverse effects, alternatives, short- and long-term commitments, and any irreversible alteration of resources. These statements must also include the comments of state and local environmental agencies, as well as those federal agencies with expertise in environmental matters. The Council on Environmental Quality Interim Guidelines for Preparation of Environmental Statements require each federal agency to establish internal procedures for implementing the act. The courts have consistently upheld the intent of the statutory national policy. (Sloan, 1971, pp. 1-4.)

Two spin-offs from the avalanche of research and reports resulting from the NEPA legislation were increased involvement of the states in federal decision making and increased awareness that the coastal zone is an area of special environmental concern. The latter result led to significant legislation that was signed into law in October 1972.

The Marine Protection Research and Sanctuaries Act of 1972 legislates control of all ocean dumping. A five-year feasibility study will determine whether it is possible to eliminate all ocean dumping. Chemical, radiological, and biological warfare agents and radioactive wastes can no longer be dumped at sea, and all other dumping will require a permit issued by the Army Corps of Engineers on the basis of criteria set by the Environmental Protection Agency. This provision means that such coastal cities as Miami, North Miami, and Miami Beach, which use ocean outfalls for sewage disposal, will be required to obtain permits under the National Pollution Discharge Elimination System (NPDES) as established by FWPCA.

Though congressional enactments dealing with local waters are too numerous to list, it is important to note that much of the thrust of recent legislation is directed toward environmental protection and conservation of natural resources (Cowan, 1973). Of major significance in this connection is the Coastal Zone Management Act of 1972, which calls for close cooperation between the federal government and state and local entities with regard to conservation and management of coastal regions. This act will have considerable impact upon local land use decisions once programs are developed; it calls for the development of comprehensive state coastal zone management plans in all participating states. Once completed, all federally funded programs must comply with the coastal zone management plan for a given area except in matters of overriding national concern as determined by the president upon recommendation of the secretary of commerce. The possible impact of the program upon questions concerning local land and water use is potentially very significant.

It needs to be emphasized that paramount federal control over local lands and waters does not of itself preclude the exercise of state and local jurisdiction in these areas. Except for this subordination to federal authority, there is no question of the constitutional right of states to regulate all land and water areas within their territorial limits. In fact, the need for concurrent state participation is frequently expressed in federal enactments, such as those in the following list, adapted from Cowan, 1973.

Major Federal Enactments Applicable to Local Waters

(1) Refuse Act (Rivers and Harbors Act of 1899)
(2) Fish and Wildlife Coordination Act of 1958
(3) National Environmental Policy Act of 1969
(4) Water Quality Improvement Act of 1970 (including 1972 amendments)
(5) Federal Boat Safety Act of 1971
(6) Coastal Management Act of 1972
(7) National Sea Grant College and Program Act of 1966 (Provides for education and training, applied research and advisory services supportive of coastal management)
(8) Federal Water Pollution Control Act of 1972, as amended (P.L. 92-500)

Major Administrative Agencies

(1) U.S. Coast Guard
(2) Army Corps of Engineers
(3) Environmental Protection Agency
(4) U.S. Fish and Wildlife Service
(5) Federal Power Commission
(6) U.S. Atomic Energy Commission
(7) Bureau of Outdoor Recreation
(8) National Park Service
(9) National Marine Fisheries Service
(10) U.S. Geological Survey
(11) National Ocean Survey (Previously known as U.S. Coast and Geodetic Survey)

As a result of new legislation and some expansion of regulatory authority through broader administrative rules, federal and state agencies are beginning to concern themselves with land use questions. At the federal level, the Army Corps of Engineers may withhold permits for a proposed waterfront development if the development is judged to have significant adverse environmental impact. Similarly, the Environmental Protection Agency has the authority to approve or disapprove local waste water treatment facilities through the

FWPCA of 1972. The consequences of this authority are significant with respect to local land use decisions.

Perhaps even more significant is the authority vested in the Department of Housing and Urban Development by virtue of the Federal Flood Disaster Protection Act. Under the provisions of this act, all conventional sources of construction financing—Federal Housing Administration (FHA), Veterans Administration (VA), or Federal Deposit Insurance Corporation (FDIC) insured funds, etc.—will be withheld unless a municipality or county adopts building codes and minimum ground floor elevations for new construction that provide adequate protection in designated flood prone areas. Regulation such as this can have the effect of stringently controlling almost all land uses in flood prone areas.

The federal government plays an increasing role in state and local decisions affecting land use. Existing laws, such as the Commerce Clause of the Constitution, have been greatly supplemented by recent legislation, such as the 1969 NEPA and the 1972 FWPCA, which provide an impetus for comprehensive planning and use of resources. Many federal grant-in-aid programs require extensive planning as a prerequisite to funding local projects. Chief sources of the federal planning grants have been Section 701 of the Housing Act of 1954, as amended, and transportation planning grants under the Federal Highway Act of 1962, and most recently the Urban Mass Transportation Act of 1964, as amended, and the Housing and Community Development Act of 1974. In the past twenty-five years almost every metropolitan area or center and many smaller cities in the United States engaged in some type of city or metropolitan planning, paid for, at least in part, with federal funds. This stimulus to planning has created a new focus in the minds of many local and state officials on the need for better management and better decision-making techniques (Clawson, 1973). This planning effort initiated a land use inventory of existing and proposed uses and of resource distribution and capability for sustaining urbanization. Policies and programs of the Department of Agriculture, the Soils Conservation Service, and the National Forest Service were examined by researchers seeking more comprehensive guidelines for planning the compatible use of resources with respect to human activities.

To many of us, it appears likely that some version of national land use planning legislation will soon become law. Until an act is passed, however, any attempt to analyze how it will operate is fruitless. Bills now pending, although differing in important respects, would make the states the chief planning agencies with financial help from the federal government. Those states that do not carry out land use planning would be penalized, probably in denial of federal assistance for particular state and local programs.

State Level

As repositories of police powers (U.S. Constitution, Amendment 10), states may regulate an act in the name of "health, safety, morals, and general welfare." These powers form the legal basis for pollution control, regulation of beaches and setback requirements, preservation of historic and wilderness sites, the power of eminent domain, etc. Traditionally, the state has delegated its land use powers to local governments, but now a move to recapture these powers is underway since many local governments have been unable to adequately respond to land use decisions that have impact beyond their political boundaries (Graham, 1972). Because states do not have sufficient funds to control land by outright purchase, the state may be expected to turn to increased use of police powers and to take a more active role in the decisions on zoning and land use in the coastal zone (O'Connor, 1972).

In recent years citizens of Florida and their elected officials have become concerned for the quality of the environment in the state. This concern has ultimately been recognized through various legislative and administrative processes. Some of the more prominent official expressions of public concern for environmental protection are embodied within the following:

(1) Bulkhead Act of 1957—provides for the establishment of a bulkhead line beyond which no further filling may occur

(2) Randall-Thomas Act of 1967—requires dredges and fill permits and an environmental assessment of the impact

(3) Amendments to the Internal Improvement Trust Fund Statute (Chap. 253, F.S.)

(4) The Florida Air and Water Pollution Control Act (Chap. 403, F.S.)

(5) The Florida Archives and History Act (Chap. 267, F.S.)

(6) An Act Creating the Coastal Coordinating Council (Sec. 370.0211, F.S.)

(7) The Florida Water Resources Act of 1972 (Chap. 373, F.S.)

(8) The Florida State Comprehensive Planning Act of 1972 (Chap. 23, F.S.)

(9) The Florida Environmental Land and Water Management Act of 1972 (Chap. 380, F.S.)

(10) The Land Conservation Act of 1972 (Chap. 380, F.S.)

(11) The Concurrent Resolution Adopting a Policy on Growth for the State of Florida, adopted by the Florida House and Senate (1974)

Taken together, these legislative expressions demonstrate a clear intent by the people of Florida and their elected representatives to achieve a quality environment (Department of Natural Resources, 1974 Draft).

The most fundamental policy guide for all state programs concerned with the natural environment is contained in Art. II, Sec. 7 of the Florida Constitution, which states: "It shall be the policy of the state to conserve and protect its natural resources and scenic beauty. Adequate provisions shall be made by law for the abatement of air and water pollution and of excessive and unnecessary noise."

The 1972 state legislature passed four landmark bills destined to reshape land use decisions of Florida. These are

(1) The Environmental Land and Water Management Act of 1972

(2) The Land Conservation Act of 1972

(3) The Water Resources Act of 1972

(4) The Florida State Comprehensive Act of 1972

These acts must be considered together because they interrelate to form a systematic mosaic of legislative and administrative strategies comprising Florida's land and water use policies (Starnes, 1974). Furthermore, two regional agencies were given wider responsibilities under these acts. The Regional Planning Agencies were designated to review Developments of Regional Impact and to act as liaison with state and federal agencies in regional planning matters (Environmental Land and Water Management Act). In southern Florida the Central and Southern Florida Flood Control District (FCD) was designated as the regional agency to effectuate policies concerning uses of waters in the Districts region (Water Resources Act). Figure 70 shows the boundaries designated for these regional agencies.

1. The Environmental Land and Water Management Act allows the state a major role in land management decisions which transcend local jurisdictional boundaries. Local governments continue to have total responsibility for land use decisions of strictly local impact. The state role, however, is to represent the public interest in decisions of regional or statewide importance. The power of the state to regulate private property does not go beyond that which is presently available to local governments. In order to determine which decisions are of sufficient importance to warrant state intervention, two designations are used, with corresponding procedures for obtaining permits.

(a) Areas of Critical State Concern. Chapter 380 of the Florida Statues established the method for designating "areas of critical state concern." The act states that the Division of State Planning "may from time to time recommend to the Administration Commission, specific areas of critical state concern." This recommendation is to include principles formulated for each such area which will guide development in a proper manner. Areas for consideration may be suggested by public agencies, private organizations, or even individual citizens. Each must be carefully considered by the division, since not more than 5 percent of the state's land (approximately 1.8 million acres) may be designated as critical under the act.

The three criteria outlined by the act for designating a critical area are that it must: (1) contain, or have a significant impact upon, environmental, historical, natural, or archaeological resources of regional or statewide importance; or (2) be significantly affected by, or have a significant effect upon, an existing or proposed major public facility or other area of major public investment; or, (3) be of major development potential as defined in the State Land Development Plan.

The recommendations of the Division of State Planning must be submitted to the Administration Commission, accompanied by the following documentation: (1) the boundaries of the proposed area; (2) the reasons why the proposed area is of critical concern to the state or region; (3) the dangers that would result from uncontrolled or inadequate development of the area; (4) the advantages that would be achieved from the development of the area in a coordinated manner; and, (5) the specific recommended principles for guiding the development of the area.

A decision must be made by the commission within 45 days of submission; those areas that are adopted become Areas of Critical State Concern. Local governments with jurisdiction over an area so designated are given six months to formulate development regulations to implement the state principles. The local government then becomes responsible for administering these regulations. In the event that a local government fails to act, the state government is required to implement the regulations itself.

Among the other significant provisions of the act is the empowerment of the state to adopt guidelines and standards for deciding whether particular land developments (such as airports, power plants, and shopping centers) are "developments of regional impact." Facilities so designated are subject to public hearings and the submission of regional impact statements whenever plans are made to build, modify, or expand them.

(b) Developments of Regional Impact. In this case, the state responds to a proposed use of such character or magnitude that it will affect the residents of "more than one county." Each regional planning agency may recommend to the Division of State Planning developments to be designated as Developments of Regional Impact (DRI). A developer may also request that the Division of State Planning indicate whether his proposal would fall under these guidelines; within 60 days the developer can expect a written, binding letter of interpretation with respect to the proposed development. Upon notice, the applicant must prepare and submit a DRI statement containing sufficient information for regional planning agency review. After public hearings, the regional planning agency prepares and submits to the local government a report and recommendation on the regional impact of the proposed development in consideration of the impact upon the region's environment and natural resources, economy, water sewer facilities, solid waste disposal, public transportation facilities, housing, and other factors that the regional agency deems important. Guidelines and standards for designating and evaluating the impact are available from the regional and state agencies. Because revisions to the guidelines are being continually up-dated, detailed information is not listed here. Examples of developments generally qualifying as DRIs are airports, attraction and recreation facilities, electrical transmission lines, mining operations, large residential developments, and shopping centers. If a development project includes more than one development of regional impact, a comprehensive impact application may be filed.

2. **Water Resources Act** provides for statewide water use planning and management in conjunction with the policies of the 1972 Environmental Land and Water Management Act and the 1972 Florida State Comprehensive Planning Act. Water use planning is a two-level effort accomplished through the establishment of state

and regional planning and management authority.

The Water Resources Act provides for control and regulation of inland and coastal waters of the state through the use of five water management districts (or agencies) that exercise the delegated powers of the Department of Natural Resources through a governing board. Each district has the responsibility to promote the conservation, development, and proper utilization of surface waters and groundwaters of the state. The Districts are required to prevent damage from flooding, soil erosion, and excessive drainage and to preserve natural resources, fish, and wildlife. Also, each District is to promote recreational development, protect public lands, and assist in maintaining the navigability of rivers and harbors. The Central and Southern Florida Flood Control District (FCD) is the management region in southern Florida (Figure 70).

Under this act the FCD shall regulate and control all ground and surface water, including the use of such water (Central and Southern Florida Flood Control District, 1974). The FCD may, for example, issue permits for the use, diversion, or withdrawal of any water in the district and for discharge into the waters in the district. The FCD may also declare that a water shortage exists within all or part of the district, and exercise its authority to ration water use or deny permits for proposed use or diversion. These jurisdictions make the FCD a powerful agency in local and state land use decisions.

3. The Comprehensive Planning Act requires the Division of State Planning to develop a comprehensive plan consisting of goals, objectives, and policies to guide the orderly social, economic, and physical growth of the state. As authorized by the act, the secretary of administration has established ten regional planning districts. Since there are ten regional planning districts and only five water management districts, the two sets of districts are not coterminous (South Florida Regional Planning Council, 1973). In South Florida, all or part of four planning regions lie within the FCD, with regions 9 and 10 covering some four-fifths of the water management district (Figure 70).

4. The Land Conservation Act passed by statewide referendum in 1972 provided for $200 million in general obligation bond funds for the state to acquire environmentally endangered lands and $40 million to acquire and improve recreation facilities. Its purpose is to meet a policy requirement for purchase of endangered lands should regulatory measures fail to meet constitutional tests. The state strategy was to implement land use planning management through regulation if possible (Starnes, 1974). If such regulation could not be reasonably imposed, the alternative is acquisition—a policy squarely in the middle of the constitutional issue of taking.

In response to the then-pending federal environmental legislation, as well as the Federal Intergovernmental Cooperation Act of 1968, the Florida Legislature in 1970 created the Coastal Coordinating Council (CCC) as the research, coordination, and planning entity in coastal zone management in Florida (Chap. 70-259, F.S.). The CCC unites in one body the directors of the three state departments with primary concern for the coastal environment, namely, the Department of Natural Resources, Department of Pollution Control, and the Trustees of the Internal Improvement Trust Fund. The Executive Director of the Department of Natural Resources serves as chairman. In its guidelines, the CCC states that the principal consideration in all coastal resource allocations will be maintenance and/or improvement of environmental quality. Public interests will be the primary criterion, and all criteria will provide for maximum retention of options for the future. The Florida Coastal Zone Master Plan will provide guidelines for regional and local planning.

The categories of use in order of priority are: (1) Preservation, with no deoverriding public interest, as determined by the governor, cabinet, and/or legislature. Ecological considerations are primary, and preservation is deemed to be of statewide significance and therefore a state-level zoning responsibility. (2) Conservation, in areas unsuited for intensive development yet usable for limited development, such as recreational use. The majority of subcategories are already set by existing state laws and cabinet policy. Conservation zoning responsibilities can be shared between state and regional or local authorities, within these limits. (3) Development, in areas well suited for intensive development and not environmentally fragile. This category does not imply actual development, but indicates that if development is to occur at all it should be in those areas so designated. Zoning for specific uses is to be a local responsibility utilizing state guidelines and specific state criteria for shoreline uses. Almost all of the southern Florida region has been mapped using these criteria (Department of Natural Resources, 1974, Draft). Accomplishments of the CCC are impressive (Coastal Coordinating Council, December 1971):

(1) Development of general guidelines for coastal zone planning.

(2) Delineation of the Florida coastal zone boundaries.

(3) Development of a management rationale through use of three basic land and water use categories: preservation, conservation, and development.

(4) Identification of the most pressing coastal zone research needs.

(5) Liaison with private coastal zone experts.

(6) Liaison with funding sources for coastal zone research.

(7) Identification of and liaison with all state and federal agencies involved with the coastal zone of Florida.

(8) Advice to regional, county, and city planning organizations in the field of coastal zone management.

(9) Compilation of state procedures for issuing permits for coastal zone activities.

(10) Information services, including a library, newsletter, resource inventory, technical bibliography.

Pollution control is primarily the responsibility of the Florida Department of Pollution Control (DPC), which is undergoing rapid transition in its program because of new federal requirements under the FWPCA of 1972. The DPC has powers analogous to those of the Federal Environmental Protection Agency. Under these statutory powers, air and water quality for the state is to be protected, maintained, and if possible upgraded. The impact of this far-reaching authority upon land use becomes obvious since the agency regulates almost all upland and coastal activities.

Along with the DPC, the Trustees of the Internal Improvement Trust Fund (TIITF) shares the major burden for managing the regulatory process. Both agencies have made extensive, continuing efforts to deal with the management of their regulatory processes: DPC primarily in response to evolving federal policy, and TIITF in response both to federal requirements and substantial changes in tradi-

Water Management Districts (July 1, 1975)

Substate Planning Districts

6 = East Central Florida Regional Planning Council
7 = Central Florida Regional Planning Council
8 = Tampa Bay Regional Planning Council
9 = Southwest Florida Regional Planning Council
10 = South Florida Regional Planning Council

FIGURE 70: Substate planning districts and water management districts.

tional state attitudes toward coastal development. The TIITF is the trustee of all public domain areas in Florida. For years it strongly encouraged development of tidelands, under conditions of the least restraint. Recently, the TIITF has introduced more stringent policies but still has a backlog of uncompleted projects which it approved in the past. However, public attitude has changed so rapidly in the past three years that it is doubtful if future applicants for grants of public tidelands will meet with favor.

Specific provision is made by the TIITF for prior local review of applications: "All permits are issued subject to compliance . . . with all local ordinances and regulations. . . ." There is also a consolidation of TIITF and DPC permit application requirements: "Application for a Trustee's permit constitutes a request for the Department of Pollution Control to consider issuance of the necessary water quality approval, in accordance with U.S. Public Law 91-224, or Chapter 403, Florida Statutes."

As the principal enforcement vehicle for achieving national water quality goals established in the 1972 FWPCA, the 402 permit program (NPDES), has vast implications for development in the coastal zone. Application of stringent water quality standards to coastal development is likely to affect the potential siting, design, construction methods, and economics for such development. Under the NPDES program, pollutants from any point source that are discharged into navigable waters require a 402 permit. Discharges are not limited to effluent from municipal and industrial sources. The act defines "pollutants" to include dredge spoil, rock, and sand; and "point source" encompasses "any discernible confined and discrete conveyance," including "any ditch, channel, conduit or discrete fissure" (Sec. 502). These definitions directly apply the substantive provisions of the act to dredging, filling, and other construction associated with many coastal development projects in Florida.

Closely associated with the major implications of the 1972 FWPCA is a recently promulgated EPA policy on wetlands (Federal Register, Vol. 38, No. 84, May 2, 1973) that would preserve "the wetland ecosystems and protect them from destruction through waste water or nonpoint source discharges, and their treatment or control, or the development of waste water treatment facilities, or by physical, chemical, or biological means." When combined with the policies and responsibilities administered by EPA under the 1972 FWPCA, this wetlands policy has particular relevance to Florida's coastal zone. Specifically, the policy calls for EPA to minimize alterations in the quality or quantity of the natural character of water inflow and withdrawals; to protect wetlands from adverse dredging and filling practices, siltation, or the addition of toxic materials arising from nonpoint source wastes and through construction activities; and to prevent violations of applicable water quality standards.

Sale of Tidelands

State jurisdiction in the coastal zone is based on control of submerged lands and on the police powers. English common law established the general principle that the lands seaward of the mean high-water mark were sovereign lands to be held in trust for the people. The case of Shively v. Bowley 152 U.S. 1 (1893) affirms operation of this principle in the United States. Sovereign land acquired with statehood and a federal grant of swamp and overflow lands in 1850 (Fla. Constitution, Art. XII, Sec. 4), as well as old Spanish grants, establish state ownership. Titles to this land are now invested in the TIITF by legislative action (F.S., Sec. 253.12, 1957). The TIITF may sell these lands if such action is in the public interest, but only to upland owners and only down to the bulkhead line, which is to be determined by local authorities. Criteria for the location of bulkhead lines, as defined in the Florida Administrative Code, specifically include marine life, natural beauty, and recreation (Fla. Admin. Code, Sec. 200-2.02, 200-2.04). These limitations were largely ignored for years, and many state lands were sold to private interests. After a moratorium was put into effect in 1967 by the state attorney general, amendments were made to the Administrative Code regarding sale of tidelands and establishment of bulkhead lines. These amendments required the TIITF to determine, with the aid of biological, ecological, and hydrological studies made under the direction of the State Conservation Board (now part of the Department of Natural Resources), what action might be in the public interest [Sec. 253.12, Subsec. (7), Supp. 1968, F.S.]. Traditionally, the public interests being protected were navigation, commerce, and fisheries, until the specific expression of legislative intent to include environmental and aesthetic values (O'Connor, 1972; Garretson, 1968; U.S. Department of Interior, 1970). It should be noted that determination of high-tide level is of special significance, since the difference of a few inches in a broad shallow tidal area can mean the gain or loss of thousands of square yards of submerged bottoms. In 1969 the legislature extended jurisdiction of the TIITF to include bottoms under navigable waters (Chap. 69-308, F.S.).

A March 15, 1974, decision by the U.S. Middle District of Florida (U.S. v. Holland, 373 F. Supp. 665,6 ERC 1388) concurred with the definition of navigability under the 1972 FWPCA which empowers the Army Corps of Engineers to protect wetlands from the damaging effects of dredge and fill above the mean high-water line. The court in Holland said (6 ERC at 1395): "The court is of the opinion that the mean high water line is no limit to Federal authority under FWPCA. While the line remains a valid demarcation for other purposes, it has no rational connection to the aquatic ecosystems which the FWPCA is intended to protect."

Local Level

Land use decisions with regard to zoning, uses, improvements, etc., are usually made at the local level. City and county commissions, acting through their zoning, building, planning, and public safety boards, regulate and determine the use of the land. Several of the tools for implementing these decisions are embodied in five major governmental powers (Coastal Coordinating Council, 1974; Hagman, 1971):

1. **Police Power.** (a) Zoning Ordinances are probably the oldest and most prevalent use of the police power (the power to regulate land) to direct land use. In 1926 the U.S. Supreme Court (one of the most conservative in American history) upheld Euclid, Ohio v. Ambler Realty Co., a precedent-making decision that empowered municipalities to zone land for various uses and to prevent discordant uses. Land use zoning has been a powerful tool in minimizing discordant land use charges and protecting property values. In some cases, however, it has been a tool for racial and economic segregation. Exclusionary by intent, zoning restricts what we may or may not do with land. In this exclusionary sense many visionary planners consider zoning to be given too much importance in the local decision-making process (McAllister, 1973). The limiting characteristics of zoning to some degree stifle imagination and change and promote segregation of the classes while encouraging monotony in the design and layout of the city. In theory, a city or regional plan should precede land use zoning action. The comprehensive plan is the goal objective, and the zoning ordinance is one method in its implementation. In practice, however, the reverse has been true almost since land use zoning began.

The most often stated purposes of zoning ordinances are listed in the Florida Statutes, Chap. 176 as follows:

(1) To lessen congestion in the streets.

(2) To secure safety from fire, panic, and other damages.

(3) To promote the health and general welfare.

(4) To prevent overbuilding of the land.

(5) To avoid undue concentration of the population.

(6) To facilitate the adequate provision of transportation, water, sewerage, schools, parks, and other requirements.

(b) Land Subdivision Regulations are a use of the police power that deals with the mechanics of platting land for future marketing purposes. The Division of Florida Land Sales at the state level has jurisdiction over any land "which is divided or is proposed to be divided for the purpose

of disposition into 50 or more lots...."
(Chap. 478, F.S.). Subdivision regulations at the local level for ten or more lots include stipulations on minimum lot sizes; required improvements, such as sidewalks; sanitary and storm sewers; dedication of land for streets; dedication of open space; etc. Some of the most perplexing urban and suburban problems—undesirable land use patterns, monotony, insufficient open space, etc.—can be directly attributed to the manner in which subdivisions are laid out.

(c) Special District Regulations are frequently used in development, protection, and administration of land. For example, in urbanizing areas special districts are used to implement management of water, sanitation, parks, historic preservation, etc. Powers of districts vary greatly but are generally quite forceful toward specific objectives and can require lot owners to pay their share of the cost to provide sewers even though the owners may plan to hold their lots as open spaces.

(d) Building, Fire, and Safety Codes are commonly used to impose standards for construction of buildings and facilities. The most frequent criticism of these codes is that they tend to be out-of-date, are inflexible, and do not adequately reflect special circumstances of the area. Performance standards may be used in a much broader fashion to also control vegetation removal, storm water runoff, canal design, etc.

2. **Eminent Domain** (the power to appropriate land for a necessary public use) provides for acquiring sites needed for public use, such as highway, street, and utility right-of-ways, parks, and open space, and other public improvements. The primary considerations on this power are that the land can only be used for public purposes and that just compensation must be paid. Eminent domain can also be applied by the state and federal government when implementing public improvements.

3. **Spending Power,** the power to purchase outright lands for public purposes, is many times the simplest and most direct means of accomplishing a land use goal, but is severely limited by available funds.

4. **Proprietary Power** subjects land that is in public ownership to far greater control than private property; therefore, there is an increasing use of proprietory power as a means of directing land use. So acquired, most of the land is sold again for private redevelopment. Restrictive covenants, however, are ordinarily imposed to ensure future compliance with the overall plan of the area. Greater use of this power, together with eminent domain and spending power, could serve as a useful tool in the implementation of policy goals.

5. **Taxation Power** provides governments with the means to collect most of the revenues necessary to finance their many operations and functions. It also provides a tool they can use for various nonfiscal and regulatory purposes, such as encouraging land use controls through tax incentives, transferring development rights, etc., as discussed more fully in the previous parts of this section.

Summary and Conclusions

Land use, controls, and individual rights are complex issues, a fact fully realized by anyone trying to unravel and understand them. In the business sense, land is considered to be a commodity to be dealt with at an exchange value. In the governmental sense, land plays a major role in decision making when solvency, growth, employment, and social and environmental issues are considered. As greater controls are put upon the land and its use, the issue of taking must be addressed. There is no clear line between regulation of land and taking of land. Over the last 50 years the state courts have decided literally hundreds of cases, each of which determined whether the value of a particular land use regulation did or did not outweigh the loss of property value to a particular landowner. An interesting trend is emerging, and although the number of cases is still small, there is a strong tendency on the part of the courts to approve land use regulations if the purpose of the regulation is statewide or regional in nature rather than merely local (Bosselman, Callis, Banta, 1973). Although the courts are also supporting local land use regulations with a reasonable degree of consistency, they show an obvious preference for regulations having broad multipurpose goals, such as restriction for the purpose of common good.

For practical purposes, land use regulation is a necessity. More and more demands placed upon fewer and fewer resources underscore this necessity. Technology, while it advances our convenience and well-being, creates new issues that must be considered, such as invasion of privacy, and protecting the public health and the standard of living. Freedom is a well-chosen word when its definition includes not only the rights of the individual to do as he pleases, but respect for the rights of others as well. Regulation of the environment in respect to social and environmental well-being will continue to increase, not because one group wishes to deny another freedoms but rather because we choose to protect our common rights and those of future generations.

Alternatives are available to the equitable resolution of land use control. The concept of regulation of land conjures up thoughts of taking, whereas, in fact, in its broadest perspective, regulation provides incentives for landowners, with government achieving its intended goals. Land use is regulated to minimize discordant land uses, manage the growth of an area subsequent to public services and facilities, and to promote the public health, safety, and general welfare. It is easy to understand how conflicts can occur between landowners and government, since each one can be motivated by different goals and objectives. The following is a summary of two techniques that could easily be applicable in some modified form to the southern Florida region.

Land Banking

Land banking is the acquiring and placing of land in a temporary holding status to be turned over for development at some future date, with the cost of the original acquisition recouped when the site is developed. The two general objectives of land banking are:

1. Public Use: To acquire and hold for future public use those lands which will serve as sites for parks, schools, utilities, low-income housing, etc. This theoretically allows the participating level of government to acquire lands at a lower market price than if it waited to purchase the land until the area was already developed when land values would be high and sites scarce.

2. Private Development: To acquire and hold those lands for resale for future development as industrial, housing, or commercial uses. This allows for resale of the land at a time complementary with planning and capital improvements objectives.

A land banking program can make land use plans more than goal statements, such as low-cost housing in the vicinity of a rapid transit station. The link between comprehensive land use planning and actual program implementation has traditionally been weak. To be an effective planning and management tool, land banking must be carried out in full coordination with both the comprehensive land use plans and the programming of public facilities and services.

Closely related to the need to coordinate development decisions is the need for government to act quickly to recognize and capture a good site before its price gets too high. Caution is necessary, of course, and while reduced tax revenue may be offset by lower-cost financing, the government will have to develop administrative mechanisms for coordinated site selection.

Land banking is a concept which has existed for many years. The Scandinavian countries, Canada, and Great Britain have used land banking for decades (Holbein, 1975). The Puerto Rico Land Administration has been operating since the 1960s. In the United States, the Housing and Community Development of 1974 provides federal guarantees for local bonds to finance property acquisition, thus giving local governments the potential financial resources. Some local governments used land banking for low-income housing. The 1972 Land Conservation Act provided $200 million in general obligation bond funds for Florida to acquire environmentally endangered lands. While this act is not fully a land banking mechanism as proposed here, its general purpose was to meet a policy requirement for purchase of endangered lands should regulatory measures fail to meet constitutional tests.

Transfer Development Rights

The concept of transferable development rights is based upon the premise that the primary factor in a real estate transaction is not the land itself but the way in which the land can be used. Separating the uses from the physical containment of the property therefore permits a more flexible process of value exchange. More importantly, in theory it releases a total commitment of the physical land from the transaction agreement, giving flexibility and equity to land ownership and regulation.

In this transfer exchange mechanism, those landowners holding land that is regulated to a low-intensity use could sell their allocated development rights (units, height, uses, etc.) to a land holder owning land regulated to a higher intensity use. The land from which the development rights were sold would stay on the tax rolls at a greatly reduced rate, while those which have increased their development potential through purchase of rights would conversely increase in tax value, roughly balancing the community's tax revenue. Such a system would not interfere with the free-market value prospectus of land transaction and marketing, and in fact would create a new dimension in land negotiations. All the landowners would have a potential benefit from the proposed intensive development, and the community would, in effect, achieve its primary goal of implementing a viable, comprehensive plan.

appendix: Normals, Means, and Extremes (Miami and Key West)

Station: **MIAMI, FLORIDA** — INTERNATIONAL AIRPORT — Standard time used: **EASTERN** — Latitude: **25° 48′ N** — Longitude: **80° 16′ W** — Elevation (ground): **7** feet — Year: **1973**

Month	Daily maximum	Daily minimum	Monthly	Record highest	Year	Record lowest	Year	Normal heating degree days (Base 65°)	Normal total	Maximum monthly	Year	Minimum monthly	Year	Maximum in 24 hrs	Year	Snow Mean total	Snow Max monthly	Year	Snow Max 24 hrs	Year	RH 01	RH 07	RH 13	RH 19	Mean speed	Prevailing direction	Fastest speed	Fastest direction	Year	Pct. sunshine	Mean sky cover	Clear	Partly cloudy	Cloudy	Precip .01"+	Snow 1.0"+	Thunderstorms	Heavy fog	90° & above	Max 32° & below	Min 32° & below	0° & below
(a)	(b)	(b)	(b)	9		9		(b)	(b)	31		31		31		31	31		31		9	9	9	9	24	15	16	16		25		24	24	24	31	24	24	25	9	9	9	9
J	75.6	58.7	67.2	86	1967	35	1971+	53	2.15	6.66	1969	0.04	1951	2.68	1973	0.0	0.0		0.0		80	84	60	70	9.3	NNW	37	31	1966		5.3	10	12	9	7	0	1	2	0	0	0	0
F	76.6	59.0	67.8	87	1971	36	1967	67	1.95	6.56	1966	0.01	1944	5.73	1966	0.0	0.0		0.0		78	82	57	66	10.1	ESE	38	30	1966		5.3	10	11	8	6	0	1	1	0	0	0	0
M	79.5	63.0	71.3	90	1971	37	1968	17	2.07	7.22	1949	0.02	1956	7.07	1949	0.0	0.0		0.0		77	82	57	65	10.4	SE	46	04	1966		5.3	9	14	8	6	0	2	1	*	0	0	0
A	82.7	67.3	75.0	96	1971	46	1971	0	3.60	10.21	1960	0.07	1971	5.18	1960	0.0	0.0		0.0		76	80	55	64	10.5	ESE	32	23	1958		5.4	8	15	7	6	0	3	1	1	0	0	0
M	85.3	70.7	78.0	93	1967	61	1971+	0	6.12	18.54	1968	0.44	1965	8.42	1958	0.0	0.0		0.0		80	82	60	70	9.4	ESE	37	32	1961		5.8	6	15	10	10	0	7	*	2	0	0	0
J	88.0	73.9	81.0	94	1967	67	1972+	0	9.00	22.36	1968	1.81	1945	7.63	1966	0.0	0.0		0.0		84	87	68	76	8.2	SE	37	13	1967+		6.9	3	14	13	15	0	13	0	6	0	0	0
J	89.1	75.5	82.3	96	1969	70	1965	0	6.91	13.51	1947	1.77	1963	4.55	1952	0.0	0.0		0.0		83	86	65	73	7.8	SE	38	18	1962		6.5	3	16	12	16	0	15	*	8	0	0	0
A	89.9	75.8	82.9	96	1970	70	1967	0	6.72	16.88	1943	1.65	1954	6.92	1964	0.0	0.0		0.0		83	87	66	74	7.6	SE	74	36	1964		6.5	2	18	11	17	0	16	*	9	0	0	0
S	88.3	75.0	81.7	93	1968+	70	1966+	0	8.74	24.40	1960	2.63	1951	7.58	1960	0.0	0.0		0.0		86	90	68	78	8.2	ESE	69	06	1965		6.8	2	15	13	17	0	11	*	5	0	0	0
O	84.6	71.0	77.8	90	1969+	56	1968	0	8.18	21.08	1952	1.50	1962	9.95	1948	0.0	0.0		0.0		84	88	66	75	9.1	ENE	41	05	1966		6.1	6	13	12	15	0	5	*	*	0	0	0
N	79.9	64.5	72.2	87	1973+	40	1970	13	2.72	13.15	1959	0.09	1970	7.93	1959	0.0	0.0		0.0		80	84	60	70	9.3	N	32	08	1966		5.2	9	13	8	8	0	1	1	0	0	0	0
D	76.6	60.0	68.3	85	1968+	34	1968	56	1.64	6.39	1958	0.13	1968+	4.38	1964	0.0	0.0		0.0		78	83	58	69	8.9	N	38	32	1967		5.2	10	12	9	7	0	1	1	0	0	0	0
YR	83.0	67.9	75.5	96	APR. 1971+	34	DEC. 1968	206	59.80	24.40	SEP. 1960	0.01	FEB. 1944	9.95	OCT. 1948	0.0	0.0		0.0		81	84	62	71	9.0	ESE	74	36	AUG. 1964		5.9	77	168	120	128	0	75	7	31	0	0	0

Means and extremes above are from existing and comparable exposures. Annual extremes have been exceeded at other sites in the locality as follows:
Airport locations: highest temperature 100 in July 1942; lowest temperature 28 in January 1940; maximum precipitation in 24 hours 12.58 in April 1942.
City locations: lowest temperature 27 in February 1917; maximum precipitation in 24 hours 15.10 in November 1925; fastest mile of wind 122 from the South in October 1950. Miami Beach (Allison Hospital): Fastest mile of wind 132 from the East, September 18, 1926.

Station: **KEY WEST, FLORIDA** — INTERNATIONAL AIRPORT — Standard time used: **EASTERN** — Latitude: **24° 33′ N** — Longitude: **81° 45′ W** — Elevation (ground): **4** feet — Year: **1973**

Month	Daily maximum	Daily minimum	Monthly	Record highest	Year	Record lowest	Year	Normal heating degree days (Base 65°)	Normal total	Maximum monthly	Year	Minimum monthly	Year	Maximum in 24 hrs	Year	Snow Mean total	Snow Max monthly	Year	Snow Max 24 hrs	Year	RH 01	RH 07	RH 13	RH 19	Mean speed	Prevailing direction	Fastest speed	Fastest direction	Year	Pct. sunshine	Mean sky cover	Clear	Partly cloudy	Cloudy	Precip .01"+	Snow 1.0"+	Thunderstorms	Heavy fog	90° & above	Max 32° & below	Min 32° & below	0° & below
(a)	(b)	(b)	(b)	21		21		(b)	(b)	25		25		25		25	25		25		22	25	25	25	20	3	25	25		15	21	21	21	21	25	25	25	25	25	25	25	25
J	75.6	65.8	70.7	85	1960+	46	1971	16	1.67	9.27	1958	0.03	1960	4.43	1953	0.0	0.0		0.0		81	83	69	77	12.1	NE	71	N	1958	70	5.1	11	12	8	7	0	1	*	0	0	0	0
F	76.6	66.5	71.6	85	1952	47	1958	75	1.85	4.46	1965	0.02	1948	2.54	1966	0.0	0.0		0.0		78	80	67	75	12.3	SE	63	WNW	1952	75	4.7	11	11	6	6	0	1	*	0	0	0	0
M	79.4	69.8	74.6	87	1949	53	1968	5	1.56	4.41	1958	T	1971	3.10	1968	0.0	0.0		0.0		77	79	66	73	12.6	SE	47	SW	1953	80	4.6	13	12	6	5	0	2	0	0	0	0	0
A	82.5	73.6	78.1	89	1965	55	1950	0	2.17	12.83	1948	0.00	1959	3.15	1950	0.0	0.0		0.0		76	77	64	71	12.7	ESE	48	NNW	1950	84	4.5	13	12	5	4	0	2	0	0	0	0	0
M	85.3	76.4	80.9	91	1953+	66	1960+	0	2.51	12.90	1960	0.00	1959	8.89	1960	0.0	0.0		0.0		76	76	65	71	11.0	ESE	51	E	1953	79	5.0	10	14	7	7	0	4	0	1	0	0	0
J	87.9	79.1	83.5	94	1952+	68	1961	0	4.55	14.43	1972	0.90	1948	4.00	1961	0.0	0.0		0.0		79	78	68	73	9.9	SE	60	SE	1966	70	6.3	4	15	11	13	0	10	0	7	0	0	0
J	89.2	80.0	84.6	95	1951	69	1952	0	4.11	11.69	1970	0.54	1961	3.05	1970	0.0	0.0		0.0		77	77	66	72	10.1	ESE	52	ESE	1952	75	6.3	4	17	10	13	0	13	0	15	0	0	0
A	89.5	79.9	84.7	95	1957	68	1952	0	4.47	11.34	1945	2.25	1969	3.23	1952	0.0	0.0		0.0		78	77	67	73	9.4	ESE	56	S	1949	76	6.2	3	18	10	14	0	14	0	16	0	0	0
S	87.8	78.6	83.2	94	1951	70	1960+	0	7.34	18.45	1963	1.70	1951	6.65	1963	0.0	0.0		0.0		79	80	69	76	10.2	ESE	122	NW	1948	67	6.6	3	16	11	16	0	11	*	7	0	0	0
O	84.0	75.2	79.6	93	1962	60	1957	0	5.57	21.57	1969	0.74	1972	8.47	1971	0.0	0.0		0.0		81	82	69	76	11.3	ENE	84	SE	1966	66	5.8	8	12	11	13	0	4	*	0	0	0	0
N	79.6	70.6	75.1	88	1964+	49	1959	0	2.67	9.01	1959	0.13	1961	7.33	1959	0.0	0.0		0.0		80	82	69	76	12.0	ENE	76	NW	1947	71	4.8	11	12	7	7	0	1	*	0	0	0	0
D	76.4	66.6	71.5	85	1961	46	1962+	18	1.52	4.84	1958	0.18	1968	4.60	1959	0.0	0.0		0.0		81	83	69	77	12.0	NE	47	N	1958	73	5.0	11	12	7	7	0	1	0	0	0	0	0
YR	82.8	73.5	78.2	95	AUG. 1957+	46	JAN. 1971+	64	39.99	21.57	OCT. 1969	0.00	APR. 1959	8.89	MAY 1960	0.0	0.0		0.0		79	80	67	74	11.3	ESE	122	NW	SEP. 1948	74	5.4	102	163	100	112	0	62	1	46	0	0	0

Means and extremes above are from existing and comparable exposures. Annual extremes have been exceeded at other sites in the locality as follows:
Highest temperature 97 in August 1956; lowest temperature 41 in January 1886; maximum monthly precipitation 23.56 in October 1933; maximum precipitation in 24 hours 19.88 in November 1954.

(a) Length of record, years, based on January data. Other months may be for more or fewer years if there have been breaks in the record.
(b) Climatological normals (1941-1970).
* Less than one half.
+ Also on earlier dates, months, or years.
T Trace, an amount too small to measure.
  Below zero temperatures are preceded by a minus sign.
‡ ≥ 70° at Alaskan stations.

The prevailing direction for wind in the Normals, Means, and Extremes table is from records through 1963.

Unless otherwise indicated, dimensional units used in this bulletin are: temperature in degrees F.; precipitation, including snowfall, in inches; wind movement in miles per hour; and relative humidity in percent. Heating degree day totals are the sums of negative departures of average daily temperatures from 65° F. Cooling degree day totals are the sums of positive departures of average daily temperatures from 65° F. Sleet was included in snowfall totals beginning with July 1948. The term "Ice pellets" includes solid grains of ice (sleet) and particles consisting of snow pellets encased in a thin layer of ice. Heavy fog reduces visibility to 1/4 mile or less.

Sky cover is expressed in a range of 0 for no clouds or obscuring phenomena to 10 for complete sky cover. The number of clear days is based on average cloudiness 0-3, partly cloudy days 4-7, and cloudy days 8-10 tenths.

& Figures instead of letters in a direction column indicate direction in tens of degrees from true North; i.e., 09 - East, 18 - South, 27 - West, 36 - North, and 00 - Calm. Resultant wind is the vector sum of wind directions and speeds divided by the number of observations. If figures appear in the direction column under "Fastest mile" the corresponding speeds are fastest observed 1-minute values.

To 8 compass points only.

SOURCE: "Local Climatological Data," U.S. Department of Commerce, National Oceanic and Atmospheric Administration, Environmental Data Service, 1973.

appendix: Normals, Means, and Extremes (Fort Myers and West Palm Beach)

Station: **FORT MYERS, FLORIDA** — PAGE FIELD — Standard time used: **EASTERN** — Latitude: 26° 35' N — Longitude: 81° 52' W — Elevation (ground): 15 feet — Year: 1973

Month	Normal Daily max	Normal Daily min	Normal Monthly	Record highest	Year	Record lowest	Year	Normal heating deg days (Base 65°)	Precip Normal total	Precip Max monthly	Year	Precip Min monthly	Year	Precip Max in 24 hrs	Year	Snow Mean total	Snow Max monthly	Year	Snow Max in 24 hrs	Year	RH 01	RH 07	RH 13	RH 19	Wind Mean speed	Prevailing dir	Fastest mile Speed	Dir	Year	Pct poss sun	Mean sky cover	Clear	Partly cloudy	Cloudy	Precip .01+	Snow 1.0+	Thunder-storms	Heavy fog	Max 90+	Max 32 below	Min 32 below	Min 0 below
(a)	(b)	(b)	(b)	14		14		(b)	(b)	34		34		34		34	34	34		34	13	13	13	13	28	10	24	24			25	32	32	32	33	33	29	30	13	13	13	13
J	74.7	52.3	63.5	88	1962	28	1966	128	1.64	6.04	1958	0.00	1950	2.25	1955	0.0	0.0		0.0		86	88	59	74	8.6	E	40	25	1958		5.0	11	11	9	5	0	1	5	0	0	*	0
F	76.0	53.3	64.7	92	1962	32	1967	125	2.03	4.65	1963	T	1944	2.60	1969	0.0	0.0		0.0		85	88	56	72	9.2	E	39	25	1958		4.9	11	10	7	6	0	1	3	*	0	*	0
M	79.7	57.3	68.5	90	1961	36	1968	48	3.06	18.58	1970	0.05	1956	7.92	1970	0.0	0.0		0.0		83	88	53	68	9.6	SW	46	35	1970+		4.9	12	11	8	5	0	2	3	*	0	0	0
A	84.8	61.8	73.3	97	1972+	44	1971	0	2.03	7.66	1941	T	1970+	3.82	1943	0.0	0.0		0.0		85	88	50	66	9.0	E	39	20	1958		4.6	11	13	6	5	0	3	2	2	0	0	0
M	89.0	66.4	77.7	98	1962	52	1971	0	3.99	10.32	1968	0.34	1962	4.57	1972	0.0	0.0		0.0		84	87	49	66	8.2	E	40	22	1965		4.9	10	14	7	8	0	8	1	13	0	0	0
J	90.5	71.7	81.1	97	1964	63	1972	0	8.89	16.10	1959	3.73	1944	6.67	1959	0.0	0.0		0.0		88	88	60	75	7.5	E	46	12	1966		6.1	5	15	10	15	0	16	*	18	0	0	0
J	91.1	73.8	82.5	97	1968+	69	1973	0	8.90	15.28	1941	2.28	1964	4.06	1965	0.0	0.0		0.0		88	88	60	75	6.9	ESE	45	18	1952		6.5	2	17	12	18	0	23	*	24	0	0	0
A	91.5	74.1	82.8	96	1968	70	1973+	0	7.72	16.22	1972	3.98	1963	6.73	1967	0.0	0.0		0.0		88	89	61	77	6.9	E	39	25	1957		6.3	3	17	11	18	0	21	*	24	0	0	0
S	89.8	73.4	81.6	95	1972+	68	1967	0	8.71	16.60	1969	2.33	1972	9.34	1962	0.0	0.0		0.0		88	90	62	78	7.8	E	92	05	1960		6.2	4	15	11	16	0	14	*	18	0	0	0
O	85.3	67.5	76.4	93	1973+	52	1973+	0	4.37	12.04	1959	0.05	1963	10.85	1951	0.0	0.0		0.0		86	88	58	74	8.5	NE	45	23	1953		5.1	11	12	8	9	0	4	4	4	0	0	0
N	79.9	58.8	69.4	90	1972	34	1970	44	1.31	3.85	1972	T	1944	3.34	1960	0.0	0.0		0.0		87	89	57	75	8.3	NE	31	27	1960		4.4	13	11	6	4	0	1	2	*	0	0	0
D	75.9	53.6	64.8	88	1972	26	1962	112	1.30	5.42	1940	0.10	1956	3.00	1969	0.0	0.0		0.0		88	89	56	75	8.3	NE	35	33	1967+		4.7	12	11	8	5	0	1	4	0	0	*	0
YR	84.0	63.7	73.9	98	MAY 1962	26	DEC. 1962	457	53.95	18.58	MAR. 1970	0.00	JAN. 1950	10.85	OCT. 1951	0.0	0.0		0.0		86	88	57	73	8.2	E	92	05	SEP. 1960		5.3	105	157	103	114	0	95	22	103	0	1	0

Means and extremes above are from existing and comparable exposures. Annual extremes have been exceeded at other sites in the locality as follows: Highest temperature 101 in July 1942; lowest temperature 24 in December 1894; maximum monthly precipitation 26.91 in June 1912; maximum precipitation in 24 hours 11.70 in June 1901. Maximum monthly snowfall T in February 1899; maximum snowfall in 24 hours T in February 1899.

Station: **WEST PALM BEACH, FLORIDA** — PALM BEACH INTERNATIONAL AP — Standard time used: **EASTERN** — Latitude: 26° 41' N — Longitude: 80° 06' W — Elevation (ground): 15 feet — Year: 1973

Month	Normal Daily max	Normal Daily min	Normal Monthly	Record highest	Year	Record lowest	Year	Normal heating deg days (Base 65°)	Precip Normal total	Precip Max monthly	Year	Precip Min monthly	Year	Precip Max in 24 hrs	Year	Snow Mean total	Snow Max monthly	Year	Snow Max in 24 hrs	Year	RH 01	RH 07	RH 13	RH 19	Wind Mean speed	Prevailing dir	Fastest mile Speed	Dir	Year	Pct poss sun	Mean sky cover	Clear	Partly cloudy	Cloudy	Precip .01+	Snow 1.0+	Thunder-storms	Heavy fog	Max 90+	Max 32 below	Min 32 below	Min 0 below
(a)	(b)	(b)	(b)	9		9		(b)	(b)	35		35		35		28	28		28		9	9	9	9	31	14	24	24			25	28	28	28	31	31	31	31	9	9	9	9
J	75.0	55.9	65.5	87	1973	29	1970	83	2.60	7.92	1958	0.22	1960	6.36	1957	0.0	0.0		0.0		80	83	59	73	9.9	NW	48	29	1955		5.8	7	12	12	7	0	1	2	0	0	*	0
F	76.0	56.2	66.1	87	1964	35	1967	91	2.60	6.88	1966	0.29	1948	4.70	1966	0.0	0.0		0.0		79	81	56	69	10.4	SE	46	29	1956		5.7	8	10	10	7	0	1	1	0	0	*	0
M	79.3	60.2	69.8	92	1965	31	1968	25	3.32	11.95	1970	0.33	1970	4.88	1970	0.0	0.0		0.0		76	79	53	66	10.7	SE	51	27	1957		5.6	8	13	10	8	0	1	0	0	0	0	0
A	82.9	64.9	73.9	99	1971	45	1971	0	3.51	18.26	1942	0.04	1967	15.23	1942	0.0	0.0		0.0		74	77	52	64	10.8	E	55	32	1958		5.6	8	13	9	7	0	2	1	*	0	0	0
M	86.1	68.9	77.5	96	1971+	62	1971+	0	5.17	14.10	1904	0.39	1967	7.04	1958	0.0	0.0		0.0		76	76	57	68	9.6	ESE	45	27	1954		5.5	8	13	9	7	0	4	1	3	0	0	0
J	88.3	72.7	80.5	96	1972	62	1965	0	8.14	17.91	1966	1.07	1952	9.21	1945	0.0	0.0		0.0		83	82	67	76	8.1	ESE	71	09	1957		6.9	4	12	14	11	0	7	*	4	0	0	0
J	89.6	74.1	81.9	96	1969	68	1965	0	6.52	17.74	1941	1.22	1961	5.83	1972	0.0	0.0		0.0		84	84	65	75	7.5	ESE	45	20	1950		6.7	4	13	14	15	0	16	*	15	0	0	.0
A	90.2	74.4	82.3	97	1970	68	1969	0	6.91	13.52	1950	2.16	1959	5.89	1949	0.0	0.0		0.0		83	84	63	74	7.4	ESE	86	13	1964		6.6	3	16	12	16	0	16	*	18	0	0	0
S	88.3	74.7	81.5	93	1972	68	1973+	0	9.85	24.86	1960	2.73	1939	8.71	1960	0.0	0.0		0.0		83	85	65	76	8.7	ESE	55	03	1965		7.0	3	13	14	17	0	11	*	9	0	0	0
O	84.3	70.1	77.2	93	1971	46	1968	0	8.75	18.74	1965	1.20	1972	9.58	1965	0.0	0.0		0.0		80	83	59	72	9.9	ENE	74	16	1964		6.3	6	13	12	14	0	5	2	0	0	0	0
N	79.5	62.5	71.0	89	1973	37	1970	22	2.48	10.77	1971	0.23	1970	5.52	1972	0.0	0.0		0.0		78	81	57	71	9.9	ENE	35	34	1959		5.6	7	14	9	8	0	1	1	0	0	0	0
D	76.1	57.4	66.8	87	1972	33	1968	78	2.21	8.73	1949	0.06	1968	5.26	1955	0.0	0.0		0.0		76	78	56	71	9.8	NNW	36	07	1958		5.5	9	12	10	7	0	1	0	0	0	0	0
YR	83.0	66.0	74.5	99	APR. 1971	29	JAN. 1970	299	62.06	24.86	SEP. 1960	0.04	APR. 1967	15.23	APR. 1942	0.0	0.0		0.0		79	81	59	71	9.4	ESE	86	13	AUG. 1964		6.1	75	153	137	131	0	78	8	59	0	1	0

Ø For period February 1964 through the current year.
Means and extremes above are from existing and comparable exposures. Annual extremes have been exceeded at other sites in the locality as follows: Highest temperature 101° in July 1942.

(a) Length of record, years, based on January data. Other months may be for more or fewer years if there have been breaks in the record.
(b) Climatological normals (1941-1970).
* Less than one half.
+ Also on earlier dates, months, or years.
T Trace, an amount too small to measure.
- Below zero temperatures are preceded by a minus sign.
‡ ≥70° at Alaskan stations.

The prevailing direction for wind in the Normals, Means, and Extremes table is from records through 1963.

Unless otherwise indicated, dimensional units used in this bulletin are: temperature in degrees F.; precipitation, including snowfall, in inches; wind movement in miles per hour; and relative humidity in percent. Heating degree day totals are the sums of negative departures of average daily temperatures from 65° F. Cooling degree day totals are the sums of positive departures of average daily temperatures from 65° F. Sleet was included in snowfall totals beginning with July 1948. The term "Ice pellets" includes solid grains of ice (sleet) and particles consisting of snow pellets encased in a thin layer of ice. Heavy fog reduces visibility to 1/4 mile or less.

Sky cover is expressed in a range of 0 for no clouds or obscuring phenomena to 10 for complete sky cover. The number of clear days is based on average cloudiness 0-3, partly cloudy days 4-7, and cloudy days 8-10 tenths.

& Figures instead of letters in a direction column indicate direction in tens of degrees from true North; i.e., 09 - East, 18 - South, 27 - West, 36 - North, and 00 - Calm. Resultant wind is the vector sum of wind directions and speeds divided by the number of observations. If figures appear in the direction column under "Fastest mile" the corresponding speeds are fastest observed 1-minute values.

SOURCE: "Local Climatological Data," U.S. Department of Commerce, National Oceanic and Atmospheric Administration, Environmental Data Service, 1973.

glossary

adsorption: adhesion, in an extremely thin layer, of the molecules of gases, liquids, or dissolved substances onto the surface of solid particles, e.g., soil particles.

anaerobic: lacking in oxygen, or living in an environment lacking free oxygen.

Anastasia formation: limestone formation found in western Collier and Lee counties, eastward along the Caloosahatchee River and eastern Palm Beach County.

aquiclude: an impermeable geologic formation that confines water in an adjacent aquifer.

aquifer: a water-bearing structure of earth, gravel, or porous rock.

backpumping: the diversion of surplus freshwater landward into impoundments for storage.

barrier reef (fringing reef): long, narrow ridge of coral or rock parallel to and relatively near a coastline.

biochemical oxygen demand (BOD): a measure of the amount of oxygen consumed in the biological processes that break down organic matter in the water.

biota: animal and plant life of a region.

brackish: mixed freshwater and seawater; salt water of lower salinity than that of seawater.

Bryozoan limestone: limestone derived from the consolidated exoskeletons of tiny colonial marine animals.

Buckingham marl: clayey marl containing a considerable amount of phosphate and pebbles.

bulkhead: seawall

calcite: calcium carbonate ($CaCO_3$).

Caloosahatchee marl: gray, sandy to calcareous shell marl.

climax community: final stage of an ecological succession, e.g., hardwood forests. **See also** ecological succession.

community: group of organisms living as a closely interacting system.

coquina limestone: geologic formation composed of consolidated shell fragments.

demography: statistical study of the characteristics of human populations.

Development of Regional Impact (DRI): under the Environmental Land and Water Management Act of the State of Florida, any development, which, because of its character, magnitude, or location, affects the citizens of more than one county.

dunes: ridges or mounds of loose wind-blown material, usually sand.

ecological succession: evolution of a biological community resulting from competing organisms affecting each other and the environment; characterized by stages ranging from pioneer communities to climax communities.

ecology: a science dealing with the patterns of relations between organisms and their environment.

ecosystem: a community and its nonliving environment.

epiphyte: plant growing on another plant for support only, deriving no nutrition from the supporting plant.

erosion: wearing away of the land surface by such natural forces as wind, water, temperature changes, etc.

estuary: coastal area where freshwater and seawater mix.

eutrophication: overenrichment (overfertilization) of a body of water characterized by algae blooms, low levels of dissolved oxygen, and frequent fish kills.

evapotranspiration: process by which water vapor enters the atmosphere; a combination of evaporation from bodies of water and transpiration by plants.

exotic: introduced foreign plant or animal.

fauna: animal species of a region.

flora: plant species of a region.

foot-candle: measure of light intensity equal to the heat reflected on a surface 1 foot away from a candle.

fringing reef. See barrier reef.

greenbelt: an expanse of parkways, parks, farmlands, or other vegetated open space that encircles a town, community, or neighborhood.

hammock: dense growth of broad-leaved trees on a slightly elevated area.

head: hammock dominated by hardwood species of the temperate zone.

hydrologic cycle: sequence in which water enters the atmosphere by evaporation and transpiration, then leaves the atmosphere by precipitation, only to re-enter the atmosphere again in a perpetual cycle.

hydroperiod: period of time a soil is waterlogged.

infrastructure: physical installations necessary to support a community, (roads, power lines, rail lines, etc.)

interglacial period: time of relatively warm climate and high sea levels between two periods of glaciation.

Iowan glacial substage: geological period of the Pleistocene epoch during which sea level fell to 60 feet below present level, exposing the Floridan Plateau.

Key Largo limestone: fossil coral reef limestone formation composed of the exoskeletons of coral polyps.

Lake Flirt marl: impervious salt-fresh-water marl.

Late Wisconsin: geological period of the Pleistocene epoch during which the sea receded to 25 feet below present level.

leaching: removal by solution of the more soluble minerals from an upper soil layer to a lower layer by the action of percolating water.

littoral zone: the edge of a water body where the water is shallow enough to permit light penetration to its bottom, typically occupied by rooted plants.

marl: finely textured calcareous mud, which may become solidified into rock.

marsh: a saltwater or freshwater wetland characterized by grasses and sedges, where few, if any, trees and shrubs grow.

mean high water (MHW): the height of tidal high water averaged over a 19 year period.

Miami limestone: formation composed of oolitic and Bryozoan limestone.

moratorium: legally authorized period of delay; a suspension of activity.

muck: dark, finely grained peat.

neap tide: a tide of minimum range, occurring at the first and third quarters of the moon.

oolite (oolitic limestone): geologic formation of consolidated, round, layered particles of calcium carbonate.

ooze: deposit of organic material of the consistency of liquid mud.

organic (soils): composed of carbon compounds; produced by the decay of previously living organisms.

Pamlico interglacial period: geological period of the Pleistocene epoch in which the sea level stood 25 feet higher than present, submerging most of southern Florida.

patch reef: scattered areas of coral growth located in shallow water landward of the fringing reefs.

peat: organic soils chiefly composed of decayed vegetation.

percolation: the slow seepage of water into the ground.

permeable: capable of allowing the passage of fluids.

pervious. See permeable.

pH: a measure of hydrogen-ion concentration; pH scale ranges from 1 (highly acidic) to 14 (highly alkaline)—a pH of 7 is neutral.

Pleistocene epoch: geological period immediately preceding the present.

runoff: rainwater that flows across the land surface into bodies of water.

salt front (salt line): the line of penetration of intruding seawater.

saltwater intrusion: movement of salt water inland.

Sangamon: geological period of the Pleistocene epoch in which the sea level stood 42 feet higher than present as a result of glaciers melting.

scrub forest: a type of vegetative community consisting of low-growing shrubs and small trees.

sedimentary rock: material formed underwater, composed of precipitated minerals; often containing insoluble organic remains.

sequent: one of a sequence.

sinkhole: in Florida an area where the surface of the land has subsided or collapsed as a result of the underlying limestone being dissolved.

slough: a channel of slow-moving water.

solution hole: a large, underground hole or cavern, resulting from the dissolving of limestone by acidic water. **See also** sinkhole.

spring tide: a tide greater than average, occurring around the times of the new and full moon.

storm surge: temporary elevation of water level resulting from storm conditions.

subclimax community: a stage in an ecological succession, held in relative stability as a result of periodic fires, or other arresting influences, e.g., Florida pinelands.

subtropical: nearly tropical; of, pertaining to, or designating regions bordering on the tropical zone.

surficial: of or relating to the surface.

swale: a low-lying area commonly moist or marshy; an intermittent drainageway.

swamp: wetland characterized by shrubs or trees such as maples, gums, bald cypresses, or, in South Florida, coastal areas, mangroves; not always covered by water the year round.

Talbot formation: marine sands and consolidated sandstone.

Talbot shoreline: the shoreline defined by the high waters of the Sangamon period, during which almost all southern Florida was submerged.

thorium dating: a process by which geologic age can be determined by analyzing the radioactive decay of the element thorium contained within the formation.

transition zone: the area where one community grades into another.

transverse glades: low, intermittent drainageways cutting across the coastal ridge of southeastern Florida.

tree island: an island of trees, shrubs, and herbaceous plants growing on an elevation and surrounded by vegetation of another kind.

water table: the upper limit of the portion of soil or rock wholly saturated with water; this level can be very near the surface or many feet below it.

weathering: the physical and chemical disintegration and decomposition of rocks and minerals.

biblio-graphy

Alexander, Taylor R., and Crook, Alan G. 1973. **South Florida Environmental Study.** Appendix G, Recent and Long-term Vegetation Changes and Patterns in South Florida. Part I, Preliminary Report. Ecological Report EVER-N-51. Springfield, Virginia: National Technical Information Service, U.S. Department of Commerce.

American Law Institute. 1971. "A Model Land Development Code." Tentative Drafts No. 3, No. 4, and No. 5. Philadelphia.

Appleyard, Donald, and Lintell, Mark. 1970. **Environmental Quality of City Streets.** Berkeley: Institute of Urban and Regional Development, University of California.

Ashley, Thomas J. (No date given). **Toward Balanced Urban Growth.** Newport Beach, California: Ashley Economic Services, Inc.

Ashley Economic Services and Albert R. Veri Associates. 1974. **The Economic and Environmental Impacts of Growth on the City of Boca Raton.** Vol. III, Section IX, "Environmental Analysis." Boca Raton.

Askew, Reubin O'D., et al. 1973. **Florida 10 Million: A Scenario of Florida's Future Based on Current Trends.** Tallahassee: The State of Florida.

Association of American Geographers. 1972. **Metropolitan Neighborhoods: Participation and Conflict over Change.** Resource Paper No. 16. Washington, D.C.

Bailey, James, ed. 1973. **New Towns in America: The Design and Development Process.** Compiled for the American Institute of Architects. New York: John Wiley & Sons.

Banz, George. 1970. **Elements of Urban Form.** New York: McGraw-Hill Book Company.

Barada, Wm., ed. July 1974. **Enfo Newsletter.** Winter Park, Florida: Environmental Information Center of the Florida Conservation Foundation.

Becker, Franklin D. 1974. **Design for Living: The Residents' View of Multi-Family Housing.** Final Report to the New York State Urban Development Corporation.

Bennett, J. W. 1974. "Anthropological Contributions to the Cultural Ecology and Management of Water Resources." In **Man and Water, The Social Sciences in Management of Water Resources,** edited by L. Douglas James, pp. 34-81. Lexington: The University Press of Kentucky.

Bernard, Nieman J., Jr.; Kuhlmey, Edward L.; and Hartmann, Robert R. 1974. "Conservation and Environmental Capacities." In **National Growth.** A Special Report of the American Society of Landscape Architects, Task Force on Growth, pp. 101-118. McLean, Virginia: American Society of Landscape Architects.

Birnhak, Bruce I., Crowder, John P. 1974. **An Evaluation of the Extent of Vegetation Habitat Alteration in South Florida 1943-1970.** NTIS Report PB 231-621. Washington, D.C.: National Technical Information Services, U.S. Department of Commerce.

Bishop, William M. 1973. "Water Quality Management Plan." Prepared for Palm Beach County, Draft Volumes 2 and 3.

Bohnsack, James A. 1974. "Establishing a Model for Determining an Optimum Regional Population." Miami. Mimeographed.

Bosselman, Fred; Callies, David; and Banta, John. 1973. **The Taking Issue, An Analysis of the Constitutional Limits of the Land Use Control.** Washington, D.C.: U.S. Government Printing Office.

Bradley, James T. 1973. "Climate of Florida." Climate of the States Series No. 8. Washington, D.C.: Environmental Data Service, U.S. Department of Commerce.

Brown, Lester R. 1974. **In the Human Interest.** New York: W. W. Norton & Company.

Cambridge Survey Research. 1974. "Environment and Florida Voters: Their Concerns on Growth; Energy; Pollution; Land Use Controls; Immigration Taxes; Politics and the Future of the State." Work Paper No. 7. Gainesville, Florida: University of Florida, Urban and Regional Development Center.

Carson, John W.; Rivkin, Goldie W.; and Rivkin, Malcolm D. 1973. **Community Growth and Water Resources Policy.** New York: Praeger Publications.

Carter, Luther J. 1974. **The Florida Experience: Land and Water Policy in a Growth State.** Published for Resources for the Future, Inc. Baltimore: The Johns Hopkins University Press.

Carter, Michael, et al. 1973. **Ecosystems Analysis of the Big Cypress Swamp and Estuaries.** Atlanta: U.S. Environmental Protection Agency, Region IV.

Central and Southern Florida Flood Control District. January 1972. **In Depth Report.** Vol. 1, No. 1.

Central and Southern Florida Flood Control District. 1974. **Rules of the Central and Southern Florida Flood Control District, Chapter 16CA-O.** Adopted December 14, 1973. West Palm Beach, Florida.

Chapp, James A. 1971. **New Towns and Urban Policy: Planning Metropolitan Guide.** New York: Dunellon.

Chinoy, Edna, and Chinoy, Norman. 1970. "Mathematical Foundations for Height Control Ratios: The Chinoy S.O.S. Formulas." Prepared for the Area Planning Board of the City of Hollywood, Florida. Mimeographed.

Chitty, Nicholas, and Davis, Charles W., eds. 1972. **The Effects of the Discharge of Secondarily Treated Sewage Effluent into the Everglades.** Sea Grant Special Bulletin No. 6. Coral Gables: University of Miami, Sea Grant Program.

Clark, Dr. E. E. 1973. "Utilization of Storm Water to Solve a Water Supply Problem." In Proceedings of the Storm Water Quality Seminar, Orlando, Florida, June 4, 1973.

Clark, John. 1974. **Coastal Ecosystems.** Washington, D.C.: The Conservation Foundation.

Clawson, Marion. 1972. **Suburban Lane Conversion in the United States, An Economic and Governmental Process.** Baltimore: The Johns Hopkins University Press.

Clawson, Marion. 1973. "Historical Overview of Land-Use Planning in the United States." In **Environment: A New Focus for Land-Use Planning,** edited by Donald M. McAllister, pp. 23-54. Report NSF/RA/E-74-001. Washington, D.C.: National Science Foundation.

Commission on Future of the South. 1974. **The Future of the South.** Report of the Commission on the Future of the South to the Southern Growth Policies Board. Southern Growth Policies Board.

Coomber, Nicholas H., and Biswas, Asitk. 1973. **Evaluation of Environmental Intangibles.** New York: Genera Press.

Costonis, John J. 1974. **Space Adrift: Saving Urban Landmarks Through the Chicago Plan.** Published for the National Trust for Historic Preservation. Urbana: University of Illinois Press.

Cowan, Dorian, 1974. "Who Governs Local Waters?" In **1973-74 Completion Report on Continuing Sea Grant Project R/L-5 Community Legal Problem Services - State and Local.** Ocean and Coastal Law Program. Coral Gables, Florida: University of Miami.

Craighead, Frank C. 1971. **The Trees of South Florida.** Vol. 1. Coral Gables, Florida: University of Miami Press.

Davis, John H., Jr. 1943. **The Natural Features of Southern Florida.** Geological Bulletin No. 25. Tallahassee: State of Florida Department of Conservation, Florida Geological Survey.

Davis, John H., Jr. 1946. **The Peat Deposits of Florida.** Geological Bulletin No. 30. Tallahassee: State of Florida Department of Conservation, Florida Geological Survey.

Department of Natural Resources, Division of Recreation and Parks, Bureau of Plans, Programs, and Services. November 15, 1975. "Executive Summary of the Florida Environmentally Endangered Lands." Tallahassee. Mimeographed.

Department of Natural Resources. 1973. **Environmental Permitting.** Tallahassee.

Dickert, Thomas G., ed., with Domeny, Katherine R. 1974. **Environmental Impact Assessment: Guidelines and Commentary.** Berkeley: University of California, University Extension.

Douglas, Marjory Stoneman. 1947. **The Everglades River of Grass.** Miami Beach: Atlantic Printer and Lithographers.

Enos, P. 1970. **Carbonate Sediment Accumulations of the South Florida Shelf Margin.** Shell Development Company, Technical Program Report EPR 29-70-F.

Environmental Protection Agency. 1973. **The Quality of Life Concept: A Potential Tool for Decision Makers.** Washington, D.C.

Farb, Peter. 1968. **Face of North America.** New York: Harper Colophon Books.

Florida Defenders of the Environment. 1971. **Environmental Impact of the Cross Florida Barge Canal, with Special Emphasis on the Oklawaha Regional Ecosystem.** Gainesville.

Florida Department of Administration, Division of State Planning, Bureau of Land Planning. October 1973. **Final Report and Recommendations for the Big Cypress Area of Critical State Concern.** Tallahassee.

Florida Department of Administration, Division of State Planning, Bureau of Land and Water Management. **South Dade County, Area of Critical State Concern.** Tallahassee.

Florida Department of Natural Resources, Coastal Coordinating Council. 1973(a). **Recommendations for Development Activities in Florida's Coastal Zone.** Tallahassee.

Florida Department of Natural Resources, Coastal Coordinating Council. 1973(b). **Statistical Inventory of Key Biophysical Elements in Florida's Coastal Zone.** Tallahassee.

Florida Department of Natural Resources, Coastal Coordinating Council. 1974. **Florida Keys Coastal Zone Management Study.** Tallahassee.

Florida Power and Light Company. 1974. **1973 Annual Report from the People of Florida Power and Light Company.** Miami.

Gantz, Charlotte O. 1971. **A Naturalist in Southern Florida.** Coral Gables, Florida: University of Miami Press.

Garretson, Albert. 1968. **The Land-Sea Interface of the Coastal Zone of the United States: Legal Problems Arising out of Multiple Use and Conflicts of Private and Public Rights and Interest.** Report to the National Council on Marine Resources and Engineering Development. New York: New York University.

George, Jean Craighead. 1972. **Everglades Wildguide.** Natural History Series. National Park Service, U.S. Department of the Interior. Washington, D.C.: U.S. Government Printing Office.

Gerrish, Harold P. 1972(a). "Air Pollution in Tropical Florida." **Miami Interaction** 4(1). Coral Gables, Florida: University of Miami, Division of Continuing Education.

Gerrish, Harold P. 1972(b). Satellite and Radar Analysis of Mesoscale Weather Systems in the Tropics. Final Report. Fort Monmouth, New Jersey: Atmospheric Sciences Laboratory, U.S. Army Electronics Command.

Gerrish, Harold P., and Hiser, H. W. 1965. Mesoscale Studies of Instability Patterns and Winds in the Tropics. Final Report No. 8. University of Miami, Radar Meteorology Laboratory.

Golemon, Harry A. 1974. **Financing Real Estate Development.** Prepared for the American Institute of Architects. Englewood, N.J.: Aloray Publisher.

Good, David A. 1971. **Cost-Benefit and Cost Effectiveness Analysis: Their Application to Urban Public Services and Facilities.** Philadelphia: Regional Science Research Institute.

Gosselink, James G.; Odum, Eugene P.; and Pope, R. M. 1973. "The Value of the Tidal Marsh." Work Paper No. 3. Gainesville: University of Florida, Urban and Regional Development Center.

Graham, Senator Robert D. 1972. A Quiet Revolution—Florida's Future on Trial." **The Florida Naturalist,** October 1972, pp. 146-51.

Greely and Hansen, Sanitation Engineers, and Connell Associates, Inc., Consulting Engineers. 1973. **Water Quality Management Plan for Metro Dade County.** Prepared for Metro Dade County Planning Department. Miami.

Gruen Gruen and Associates. 1972. **The Impacts of Growth: An Analytical Framework and Fiscal Example.** Berkeley: California Better Housing Foundation, Inc.

Hagman, Donald G. 1973. **Public Planning and Control of Urban and Land Development: Cases and Materials.** St. Paul, Minn.: West Publishing Co.

Hagman, Donald G. 1971. **Urban Planning and Land Development Control Law.** St. Paul, Minn.: West Publishing Co.

Harbor of Port Everglades, Florida. 1974. **1973 Annual Report.** Hollywood-Fort Lauderdale, Florida.

Hartwell, James H., et al. 1973. **Water: Implications of the 1971 Drought Upon Dade County's Resources and Management Policies.** Miami: Mangrove Chapter, Izaak Walton League.

Hartwell, James H. 1975. Hydrologist. Personal communication.

Heald, Eric J. 1971. **The Production of Organic Detritus in a South Florida Estuary.** Sea Grant Technical Bulletin No. 6. Coral Gables, Florida: University of Miami, Sea Grant Program.

Hemenway, Gail D. 1973. **Developer's Handbook: Environmental Impact Statement.** Berkeley: Associated Home Builders of the Greater East Bay, Inc.

Hoffmeister, John E. 1974. **Land From the Sea.** Coral Gables, Florida: University of Miami Press.

Holbein, Mary Elizabeth. 1975. "Land Banking: Saving for a Rainy Day." **Planning: The ASPO Magazine** 41(1).

Holzheimer, Terry. 1975. "In Defense of Dade's Master Plan." Miami. Mimeographed (Prepublication draft).

Jackson, Daniel F. 1974. **The Ecology and Economics of Real-Estate Lakes: Phase I Overview.** Miami: Florida International University, Division of Environmental Technology and Urban Systems.

James, L. Douglas, ed. 1974. **Man and Water, The Social Sciences in Management of Water Resources.** Lexington: The University Press of Kentucky.

Jenna, William W., Jr. 1972. **Metropolitan Miami: A Demographic Overview.** Coral Gables, Florida: University of Miami Press.

Jenna, William W., Jr. 1973. **Metropolitan Broward: A Demographic Overview.** Coral Gables, Florida: University of Miami Press.

Ketchum, Bostwick H. 1972. **The Water's Edge. Critical Problems of the Coastal Zone.** Cambridge, Mass.: The M.I.T. Press.

Klein, Howard, et al. 1970. **Some Hydrologic and Biologic Aspects of the Big Cypress Swamp Drainage Area, Southern Florida.** Open file report 70003. Tallahassee: U.S. Department of Interior, U.S. Geological Survey.

Klein, Howard. 1972. **The Shallow Aquifer of Southwest Florida.** Map Series No. 53. Prepared by the U.S. Geological Survey in cooperation with the Bureau of Geology. Tallahassee: Florida Department of Natural Resources and Collier County.

Klein, Howard; Schroeder, M. C.; and Lichter, W. F. 1964. **Geology and Ground Water Resources of Glades and Hendry Counties.** Report of Investigation No. 37. Prepared by the U.S. Geological Survey in cooperation with Florida Geological Survey and the Central and Southern Florida Flood Control District. Tallahassee: State of Florida State Board of Conservation, Division of Geology, Florida Geological Survey.

Kofoed, Jack. 1960. **The Florida Story.** Garden City, New York: Doubleday and Company, Inc.

Kohout, F. A., and Hartwell, J. H. 1967. **Hydrologic Effects of Area B Flood Control Plan on Urbanization of Dade County, Florida.** Report of Investigation No. 47. Prepared by the U.S. Geological Survey in cooperation with the Central and Southern Florida Flood Control District and the Florida Bureau of Geology. Tallahassee: State of Florida Board of Conservation, Division of Geology, Florida Geological Survey.

Kreitman, Abe; Walker, Richard H.; and Beck, James A. 1974. **Water Consumption Trends within the Central and Southern Florida Flood Control Dis-** trict. Technical Publication No. 74-3. West Palm Beach: Central and Southern Florida Flood Control District.

Landrum, Ney C. 1971. **Outdoor Recreation in Florida: A Comprehensive Program for Meeting Florida's Outdoor Recreation Needs.** Tallahassee: State of Florida Department of Natural Resources, Division of Recreation and Parks.

LaRoe, Edward T. 1974. **Environmental Considerations for Water Management District 6 of Collier County—Rookery Bay Land Use Studies—Study No. 8.** Washington, D.C.: The Conservation Foundation.

Leach, S. P.; Klein, Howard; and Hampton, E. R. 1972. **Hydrologic Effects of Water Control and Management of Southeastern Florida.** Report of Investigation No. 60. Prepared by the U.S. Geological Survey in cooperation with Florida Department of Natural Resources. Tallahassee: Florida Department of Natural Resources.

Lee, Thomas N., and Yokel, Bernard J. 1973. **Hydrography and Beach Dynamics—Rookery Bay Land Use Studies—Study No. 4.** Washington, D.C.: The Conservation Foundation.

Leopold, Luna B. 1974. **Water: A Primer.** San Francisco: W. H. Freeman and Company.

Lewis, Philip H., Jr. 1972. "An Earth Aid Program: Systems and Methods for Land Resource Policy Development." Prepared for the National Conference on Land Policy, Hamisburg, Pa., June 29-30, 1972. Mimeographed.

Little, John A.; Schneider, R. F.; and Carroll, B. J. May 1970. **A Synoptic Survey of the Limnological Characteristics of the Big Cypress Swamp, Florida.** Washington, D.C.: U.S. Department of Interior, Federal Water Quality Administration, Southeast Region, Southeast Water Laboratory, Technical Services Program.

Long, Robert W., and Lakela, Olga. 1971. **A Flora of Tropical Florida.** Coral Gables, Florida: University of Miami Press.

Lugo, A. E.; Snedaker, S. C.; Bayley, S.; and Odum, H. T. 1971. **Models for Planning and Research for the South Florida Environmental Study.** University of Florida, Center for Aquatic Studies. Gainesville: National Park Service and U.S. Department of Interior.

Lugo, Ariel E., and Snedaker, Samuel C. 1974. "The Ecology of Mangroves." **Annual Review of Ecology and Systematics,** Vol. 5.

McAllister, Donald M., ed. 1973. **Environment: A New Focus for Land Use Planning.** RANN—Research Applied to National Needs, Report NSF/RA/E-74-001. Washington, D.C.: National Science Foundation.

McCluney, William R. 1971. **The Environmental Destruction of South Florida.** Coral Gables, Florida: University of Miami Press.

McGrath, Dorn C., Jr. 1972. "Population Growth and Change: Implications for Planning." In **Aspects of Population Growth Policy,** Vol. 6, pp. 397-428. Report of the Commission on Population Growth and the American Future. Washington, D.C.: U.S. Government Printing Office.

McHarg, Ian L. 1969. **Design With Nature.** Garden City, New York: The Natural History Press.

Mandelker, Daniel R. 1961. "Notes from the English: Compensation in Town and Country Planning." In 49 **California Law Review** 699, 736-37, 1961.

Marshall, Arthur R. 1971. "Repairing the Florida Everglades Basin." University of Miami Center for Urban and Regional Studies, Division of Applied Ecology. Mimeographed.

Marshall, Arthur R. 1972. "South Florida, A Case Study in Carrying Capacity." Speech before the AAAS Annual Meet- ing, Washington, D.C. Miami: University of Miami Center for Urban and Regional Studies. Mimeographed.

Marshall, Arthur R., et al. 1972. "The Kissimmee-Okeechobee Basin." A Report to the Florida Cabinet, Tallahassee, Florida. University of Miami Center for Urban and Regional Studies, Division of Applied Ecology.

Marshall, Arthur R., et al. 1973. "Project: Man and Nature, An Interdisciplinary Program to Determine the Carrying Capacity of Florida." Sponsored by the Florida Power Industry. Unpublished.

Menchik, Mark D. 1971. **Residential Environmental Preferences and Choice: Some Preliminary Empirical Results Relevant to Urban Form.** Philadelphia: Regional Science Research Institute.

Metropolitan Dade County Aviation Department. 1974. **1973 Annual Report.** Miami, Florida.

Meyer, F. W. 1971. "Preliminary Evaluation of the Hydrologic Effects of Implementing Water and Sewage Plans, Dade County, Florida." U.S. Geological Survey Open File Report 71003. Tallahassee: U.S. Geological Survey.

Miami Federal Executive Board, U.S. Office of Management and Budget. 1973. "Evacuation of Coastal Residents During Hurricanes: A Pilot Study for Dade County, Florida." Washington, D.C.

Mierav, Ronald. 1974. **Supplemental Water Use in the Everglades Agricultural Area.** Technical Publication No. 74-4. West Palm Beach: Central and Southern Florida Flood Control District.

Monroe, Frederick F. 1975. Oceanographic consultant. Personal communication.

National Flood Insurers Association. 1974. **Flood Insurance Manual, National Flood Insurance Program.** New York.

National Marine Fisheries Service. 1975. Mr. Elmer Allen, personal communication.

National Water Commission. 1973. **Water Policies for the Future.** Final Report to the President and to the Congress of the United States by the National Water Commission. Washington, D.C.: U.S. Government Printing Office.

New York State Office of Planning Coordination. 1971. **New York State Development Plan - 1.** Albany, New York.

Nicholas, James C. October 1973. "South Florida During the 20th Century." **Florida Environmental and Urban Issues** 1(1): 2-5, 14-16.

Nicholas, James C. 1975. Associate Professor, Department of Economics, Florida Atlantic University. Interview.

O'Connor, Dennis. 1972. "Legal Aspects of Coastal Zone Management in Escambia and Santa Rosa Counties, Florida (Escarosa)." Report to the Florida Coastal Coordinating Council of the Department of Natural Resources, March 1972. Mimeographed.

Odum, H. T.; Copeland, B. J.; and McMahan, E. A., eds. June 1974. **Coastal Ecological Systems of the United States.** Vol. I. Washington, D.C.: The Conservation Foundation.

Odum, William E. 1971. **Pathways of Energy Flow in a South Florida Estuary.** Sea Grant Technical Bulletin No. 7. Coral Gables, Florida: University of Miami, Sea Grant Program.

Park, Robert, Jr., and Westoff, Charles F., eds. 1972. **Aspects of Population Growth Policy.** Vol. 6. Report of the Commission on Population Growth and the American Future. Washington, D.C.: U.S. Government Printing Office.

Parker, Garald G., and Cooke, C. Wythe. 1944. **Late Cenozoic Geology of Southern Florida, with a Discussion of the Ground Water.** Tallahassee: Florida Geological Survey.

Parker, Garald G., Ferguson, G. E., and Love, S. K. 1955. **Water Resources of Southeastern Florida.** Geological Survey Water Supply Paper 1255. Washington, D.C.: U.S. Government Printing Office.

Parker, Garald G.; Hoy, D.; and Stenens, J. C. 1943. "Further Studies of Geological Relationships Affecting Soil and Water Conservation and Use of the Everglades." **Proceedings of the Soil and Crop Science Society of Florida.** Vol. 5A, pp. 33-94.

Peck, Ralph B., and Terzaghi, Karl. 1948. **Soil Mechanics in Engineering Practice.** New York: John Wiley & Sons, Inc.

Perloff, Harvey. 1963. **How a Region Grows: Area Development in the U.S. Economy.** New York: Committee for Economic Development.

Pitt, William A. J., Jr. 1974. **Effects of Septic Tank Effluent on Ground-Water Quality.** U.S. Geological Survey Open File Report 74010. Tallahassee: U.S. Geological Survey.

Polakowski, Kenneth L., et al. 1974. "A Scenic Highway System: Upper Great Lakes Region." Madison: University of Wisconsin, Institute of Environmental Studies. Mimeographed.

Port of Miami. 1974. **1973 Annual Report.** Miami, Florida.

Pride, R. W. 1973. **Estimated Use of Water in Florida, 1970.** U.S. Geological Survey Information Circular No. 83. Tallahassee: U.S. Geological Survey.

Rabinowitz, C. B., and Coughlin, R. E. 1970. **Analysis of Landscape Characteristics Relevant to Preference.** Philadelphia: Regional Science Research Institute.

Rand McNally. 1973. **Guide to Florida.** Chicago.

Real Estate Research Corporation. 1974. **The Costs of Sprawl: Environmental and Economic Costs of Alternative Residential Development Patterns at the Urban Fringe.** Detailed Cost Analysis. Washington, D.C.: U.S. Government Printing Office.

Real Estate Research Corporation. 1974. **The Costs of Sprawl: Executive Summary.** Washington, D.C.: U.S. Government Printing Office.

Reuss, Henry S. 1973. **Protecting America's Estuaries: Florida.** Parts 1-A and 1-B. House of Representatives, Conservation and National Resources Subcommittee of the Committee on Government Operations. Washington, D.C.: U.S. Government Printing Office.

Richardson, Harry W. 1973. **The Economics of Urban Size.** England: Saxon House.

Ricker, David A., and Spieker, Andree M. 1971. **Real Estate Lakes.** Geological Survey Circular 601-G. Washington, D.C.: U.S. Geological Survey.

Robas, Ann K. 1970. **South Florida's Mangrove Bordered Estuaries: Their Role in Sport and Commercial Fish Production.** Sea Grant Information Bulletin Number 4. Coral Gables, Florida: University of Miami, Sea Grant Program.

Robertson, William B. 1967. **Everglades—The Park Story.** Coral Gables, Florida: University of Miami Press.

Rockefeller Brothers Fund. 1973. **The Use of the Land: The Citizen's Policy Guide to Urban Growth.** New York: Crowell.

Rodgers, Donald P., and Crowder, John P. 1974. **Threatened Wildlife of South Florida.** South Florida Environmental Project: Ecological Report DI-SFEP-74-25. Springfield, Virginia: National Technical Information Service, U.S. Department of Commerce.

Savage, Thomas. 1972. **Florida Mangroves as Shoreline Stabilizers.** Professional Papers Series No. 19. St. Petersburg: Florida Department of Natural Resources, Marine Research Laboratory.

Scientific American, Inc. 1971. **Cities.** New York: Alfred A. Knopf.

Seymour, Whitney North, Jr., ed. 1969. **Small Urban Spaces: The Philosophy Design, Sociology and Politics of Vest Pocket Parks and other Small Urban Open Spaces.** New York: New York University Press.

Sheaffer, John R., and Zeizel, Arthur J. 1966. **The Water Resource: Planning its Use, Northeastern Illinois.** Chicago: Northeastern Illinois Planning Commission.

Sherwood, C. B.; McCoy, H. J.; and Galliher, C. F. 1973. **Water Resources of Broward County, Florida.** Report of Investigation No. 65. Prepared by the U.S. Geological Survey in cooperation with the Florida Department of Natural Resources and Broward County. Tallahassee: Department of Natural Resources.

Shevin, Robert L. 1973. Environmental Enforcement Agencies. Office of the Attorney General, Tallahassee: State of Florida.

Simpson, R. H.; Frank, N. L.; and Carrodus, Robert L. 1969. **The Hurricane Storm Surge in South Florida.** Miami: ESSA Weather Bureau.

Sloan, Irving J. 1971. **Environment and the Law.** Dobbs Ferry, New York: Oceana Publications.

Snedaker, Samuel C., and Pool, D. J. 1973. **The Role of Mangrove Ecosystems: Mangrove Forest Types and Biomass.** South Florida Environmental Project: Ecological Report DI-SFEP-74-35. Springfield, Virginia: National Technical Information Service, U.S. Department of Commerce.

South Florida Regional Planning Council. September 1973. **Water Management Chapter. Regional Guide Service.** Miami.

Starnes, Earl M. 1974. "Experiences of Florida in Planning for Land Use." Work Paper No. 6. Gainesville: University of Florida Urban and Regional Development Center.

Starnes, Earl M. 1974. "Growth Impact Statement." An Address to the American Society of Landscape Architects Task Force on National Growth, Americana Hotel, Miami Beach. Mimeographed.

Still, Homer E., Jr. February 12, 1972. "Federal-State-Local Relations Role and Functions of the State Planning and Development Clearinghouse." Enclosures in a letter to the conferees of the Fourth Mayor's Invitational Conference of Planning. Tallahassee: Florida Department of Administration, Bureau of Planning.

Strong, Ann Louise. 1971. **Planned Urban Environments. Sweden, Finland, Israel, The Netherlands, France.** Baltimore: The Johns Hopkins University Press.

Sugg, Arnold L.; Pardue, Leonard G.; and Carrodus, Robert L. 1971. **Memorable Hurricanes of the United States Since 1873.** NOAA Technical Memorandum NWS Sr-56. Fort Worth, Texas: U.S. Department of Commerce, National Oceanic and Atmospheric Administration, National Weather Service, Southern Region.

Teas, Howard J. 1974. **Mangroves of Biscayne Bay.** Miami: Metropolitan Dade County Commission.

Tebeau, Charlton W. 1973. **Past Environment from Historical Sources.** NTIS Report PB-231-711. Springfield, Virginia: National Technical Information Services, U.S. Department of Commerce.

Thompson, Ralph B., ed. August 1973. **Florida Statistical Abstract 1973.** Gainesville: University of Florida Press.

Thompson, Wilbur R. 1973. "Problems that Sprout in the Shadow of No-Growth." **Journal of the American Institute of Architects,** December 1973, pp. 30-35.

Turner, Carolyn, ed. 1973. **Managed Growth: What Cities, States, the Federal Government, and Citizen Groups are Doing to Control Growth and Change Land Use Policy.** Chicago: Urban Research Corporation.

U.S. Department of Agriculture. 1973. **Report for Kissimee-Everglades Area, Florida.** River Basin Investigation. Fort Worth, Texas: U.S. Department of Agriculture, Soil Conservation Service.

U.S. Department of Commerce, Bureau of the Census. 1969. **1969 Census of Agriculture.** Washington, D.C.: U.S. Government Printing Office.

U.S. Department of Commerce, Bureau of the Census, March 1972. **Census Tracts: Miami, Florida. Standard Metropolitan Statistical Area.** Washington, D.C.: U.S. Government Printing Office.

U.S. Department of Housing and Urban Development. 1967. **Manual on Wood Construction for Prefabricated Houses.** Washington, D.C.: U.S. Department of Housing and Urban Development, Division of International Affairs. Reprint.

U.S. Department of Interior. 1969. **Environmental Impact of The Big Cypress Swamp Jet Port.** (The Leopold Report.)

U.S. Department of the Interior. 1973. **Resource and Land Information for South Dade County, Florida.** Geological Survey Investigation I-850. Washington, D.C.: U.S. Government Printing Office.

U.S. Department of the Interior, Fish and Wildlife Service, Bureau of Sport Fisheries and Wildlife and Bureau of Commercial Fisheries. 1970. **National Estuary Study.** 7 Vols. Washington, D.C.: U.S. Government Printing Office.

U.S. Environmental Protection Agency. 1973. **Process, Proceedings, and To Control Pollution Resulting from all Construction Activity.** Washington, D.C.: U.S. Government Printing Office.

U.S. Geological Survey. 1972. **Public Water Supplies of Selected Municipalities in Florida, 1970.** Information Circular No. 81. Tallahassee.

U.S. Geological Survey. 1972. **Water Levels in Artesian and Nonartesian Aquifers in Florida 1969-70.** Information Circular No. 73. Tallahassee.

U.S. Geological Survey. 1973 to date (monthly publication). "Water Summary of South Florida." Prepared by U.S. Geological Survey in cooperation with Central and Southern Florida Flood Control District. Miami. Mimeographed.

Urban Land Institute. 1974. **Environmental Impact of Development.** Washington, D.C.: Urban Land Institute.

Urban Land Institute. 1974. **Residential Streets: Objectives, Principles, and Design Objectives.** Washington, D.C.: American Society of Civil Engineers and American Association of Home Builders.

Veri, Albert R. 1969. "Guyana Education Project: Environmental Design Criteria for a Tropical Climate. Prepared for the Government of Guyana." Mimeographed.

Veri, Albert R. 1971. **An Environmental Land Planning Study for South Dade County, Florida.** Coral Gables, Florida: University of Miami Center for Urban Studies.

Veri, Albert R. 1971. "Population Densities: How Do They Affect Environment?" **The Florida Naturalist** 44(4).

Veri, Albert R. Associates. 1972. **An Analysis of Density as it Relates to the Future Environmental Quality of Naples and Coastal Collier County.** Naples, Fla.: The Collier County Conservancy.

Veri, Albert R. Associates. 1972. **Environmental Delineation: A Guide for Land Planning and Design.** Prepared for SOCOTEC, Department de L'Etanger, Government of France.

Veri, Albert R. 1972. "Review and Evaluation of the Proposed Three Islands Development in Hallandale and Hollywood, Florida." Prepared for the governor of the state of Florida, Reubin O'D. Askew. Mimeographed.

Veri, Albert R. 1972. "The Space We Live In: Its Quality and Its Use." **Miami Interaction** 3(2). Coral Gables, Florida: University of Miami Division of Continuing Education.

Veri, Albert R., et al. 1973. **Study No. 2: The Resource Buffer Plan, A Conceptual Lane Use Study.** Water Management District No. 6, Collier County, Florida. Washington, D.C.: The Conservation Foundation.

Veri, Albert R.; Peterson, Ernest A.; and Gerrish, Harold P. 1973. "Urban Noise and Air Quality Assessment Study with Recommendation for Future Land Use and Traffic Planning." Prepared for the City of Hollywood, Florida. Unpublished.

Veri, Albert R. 1974. "Conservation and Environmental Capacity." In **National Growth.** A Special Report of the American Society of Landscape Architects, Task Force on National Growth, pp. 119-25. McLean, Virginia: American Society of Landscape Architects.

Voss, Gilbert L. 1973. "Sickness and Death In Florida's Coral Reefs." **Natural History** 82(7):41-47, August/September 1973.

White, Gilbert F. 1971. **Strategies of American Water Management.** Ann Arbor: The University of Michigan Press. Paperback.

Wilson, Susan Uhl. 1974. "Metabolism and Biology of a Blue Green Algal Mat." Master's Thesis, University of Miami, Coral Gables, Florida.

Wilson, Susan Uhl. 1975. "Biscayne Bay: Environmental and Social Systems." Special Report No. 1. Coral Gables, Florida: University of Miami, Sea Grant Program.

Wood, Roland, and Fernald, Edward A. 1974. **The New Florida Atlas: Patterns of the Sunshine State.** Tallahassee: Trend Publications, Inc.

Zim, Herbert S. 1960. **A Guide to Everglades National Park and the Nearby Florida Keys.** New York: Golden Press.